SpringerBriefs present concise summaries of cutting-edge research and practical applications across a wide spectrum of fields. Featuring compact volumes of 50–125 pages, the series covers a range of content from professional to academic.

Typical publications can be:

- A timely report of state-of-the art methods
- An introduction to or a manual for the application of mathematical or computer techniques
- A bridge between new research results, as published in journal articles
- A snapshot of a hot or emerging topic
- An in-depth case study
- A presentation of core concepts that students must understand in order to make independent contributions

SpringerBriefs are characterized by fast, global electronic dissemination, standard publishing contracts, standardized manuscript preparation and formatting guidelines, and expedited production schedules.

On the one hand, **SpringerBriefs in Applied Sciences and Technology** are devoted to the publication of fundamentals and applications within the different classical engineering disciplines as well as in interdisciplinary fields that recently emerged between these areas. On the other hand, as the boundary separating fundamental research and applied technology is more and more dissolving, this series is particularly open to trans-disciplinary topics between fundamental science and engineering.

Indexed by EI-Compendex, SCOPUS and Springerlink.

More information about this series at http://www.springer.com/series/8884

Jörg Benndorf

Closed Loop Management in Mineral Resource Extraction

Turning Online Geo-Data into Mining Intelligence

Springer

Jörg Benndorf
Department of Mine Surveying and Geodesy
University of Technology
Bergakademie Freiberg
Freiberg, Sachsen, Germany

ISSN 2191-530X   ISSN 2191-5318   (electronic)
SpringerBriefs in Applied Sciences and Technology
ISBN 978-3-030-40899-2   ISBN 978-3-030-40900-5   (eBook)
https://doi.org/10.1007/978-3-030-40900-5

This Springer imprint is published by the registered company Springer Nature Switzerland AG
The registered company address is: Gewerbestrasse 11, 6330 Cham, Switzerland

Preface

This book is written to summarize results from seven years of research in the field of closed-loop management of mineral resource extraction, in particular, applied to grade control and grade streaming.

Until 2012, the author has worked as a mine project engineer in different commodities and has been exposed to the tasks of grade control and raw materials quality management in different facets, including exploration, resource modelling, mine planning, and operational management. During his time in the industry, the first decade of the twenty-first century, a strong technical development in the application of modern sensor technology for production monitoring in mines could be observed.

Joining academia in 2012, the author's research focused on the question "How to turn production monitoring data quickly into mining intelligence?" At TU Delft, the Netherlands, the author worked together with Dr. Mike Buxton, an expert on sensor-based material characterization. Both initiated two large European projects, "Real-Time Mining" (Horizon 2020, Grant Agreement No 641989) and "RTRO-Coal" (the European Research Fund for Coal and Steel, Grant Agreement No RFCR-CT-2013-00003). During the time from 2012 to 2016, a research group lead by the author developed methods and the framework presented in this book. To acknowledge contributing researchers, Dr. Tom Wambeke, Dr. Masoud Soleymani-Shishvan, and Dr. Cansın Yüksel Pelk need to be named, who have been supported by B.Sc. and M.Sc. students Sil Roelen and Dirk-Jan Mollema. After joining TU Bergakademie Freiberg, Germany, in 2016, the research has been continued in a team around Angel Prior-Arce and supported by Paul Lorenz. The team worked on multiple industrial-scale case studies and provided implementations of the closed-loop approach on a level of proof of concept.

This book is written to summarize key results and illustrate these by examples. The text is designed neither to cover all aspects of mineral source management nor to provide a complete description of the theory. It provides an overview and a description of concepts used for an implementation of the *Closed Loop Management in Mineral Resource Extraction* approach on an introductory level. Selected technical details, new methods, examples, and some practical aspects of operational implementation are discussed. With the referenced literature, summarized at the end of each chapter, the interested reader is provided with a starting point to dive deeper into the topic.

Freiberg, Germany   Jörg Benndorf
December 2019

Contents

Chapter 1
Introduction

Abstract Digital technologies determine more and more of our day-to-day life and industrial production processes. Buzzwords, such as digital disruption or Industry 4.0 define the themes of conferences, major research programs, and technical development in many enterprises.

1.1   Towards Digitization of Mining Processes

Digital technologies determine more and more our day-to-day's life and industrial production processes. Buzzwords, such as digital disruption or Industry 4.0 define the themes of conferences, major research programs and technical development in many enterprises. These promote the use of sensor technology combined with smart data analytics and predictive modeling to derive faster relevant information in a connected world. A general goal is to provide relevant and reliable information from sensors to control options and, in this way, support an intelligent decision-making process. Our world is equipped with an enormous amount of sensors. Examples include positioning sensors, accelerometer, gyroscopes and RGB Cameras in mobile phones, distance measuring scanners in cars, satellite data and many further. It is expected that in a few years most parts of our world will be mapped or scanned by sensors several times daily. Updates about changes, detected by sensors, will be available in short time intervals, nearly in real time. Unlocking the potential of these real-time data is the current mission of global enterprises, such as Google, Amazon, or Tesla.

The question comes up, how to translate these concepts to the mineral resource extraction industry? Most mine planning and production control activities rely on geospatial deposit models as an input. The aim is typically to optimize process efficiency and mineral recovery. Underlying models are traditionally based on limited data acquired prior to extraction and consequently, the predicted production outcomes based on these models cannot always be achieved. Often, there are significant discrepancies between the model-based production forecasts of the run-off-mine ore and the actual. On the other hand, modern operations have invested into Information and Communication Technology (ICT) that delivers very rich data streams and

© The Author(s), under exclusive license to Springer Nature Switzerland AG 2020
J. Benndorf, *Closed Loop Management in Mineral Resource Extraction*,
SpringerBriefs in Applied Sciences and Technology,
https://doi.org/10.1007/978-3-030-40900-5_1

enable continuous process monitoring such as material tracking or plant performance measurement. Unlocking this potential leads to the following vision:

> Imagine your operations' planning was based on an up-to-date model, fed by a resource definition and geo-metallurgy program operating 24/7 at mine-scale...

This re-frames the mining and processing operation as a data acquisition engine, providing information for continuously updating grade control or geo-metallurgical models. The model updating does not only affect mined-out blocks but also surrounding parts of the deposit being spatially correlated, hence enabling continuously improved or even optimized operational plans. In addition, a fully updated grade control model incorporating, for example, the actual ball mill working index would grow into a rich geo-metallurgical database based not on lab-scale test work but on actual mine-scale production data. This database could be leveraged to cost-efficiently improve the geo-metallurgical model for subsequent utilization at the resource model scale and medium to long-term plans. Being able to implement such concepts in operations contributes to several step changes:

- it bridges gaps between exploration, mining, and processing by integrating and utilizing available information,
- it provides a framework for geo-metallurgical modeling based on plant data, leading to a step change in the amount of available information compared to geo-metallurgical test work from drilling campaigns,
- the newly available database on block-scale measurements rises the potential to upscale tactical geo-met grade control learnings to a resource model-scale, and
- utilizing the enormous information source of online production data will justify decreasing exploration effort upfront of the project.

Closed-loop approaches, that address this topic, have recently been developed and demonstrated to utilize up-to-date information in combination with advanced computing technology for improved production control in mineral resource extraction. Leveraging on available production data, closed-loop approaches for geo-resource extraction aim at increased resource efficiency, in terms of recovery or financial measures, using a measurement and control approach. In hydrocarbon production "closed-loop" or "real-time" approaches have received growing attention as part of various industry initiatives with names as "smart fields", "i-fields", "e-fields", "integrated operations", "closed-loop reservoir management" (CLRM), or "closed-loop field development"; see Jansen et al. (2008, 2009) and Hou et al. (2015) for further references. Independent from these developments, more recently similar concepts were proposed in solid mineral resource extraction (e.g., Benndorf et al. 2015a, b; Zogovic et al. 2015). These concepts utilize the classical concept of plan–do–check–act iterative management (Shewhart 1931). By continuously comparing model-based predictions with observations measured during production monitoring, the use of inverse modeling or data assimilation approaches can improve the model forecast for subsequent time intervals, leading to potentially new and better decisions for production control and medium-term planning. The underlying hypothesis in these approaches is

It will be possible to significantly increase life-cycle value by changing ore extraction management from a batch-type to a near-continuous model-based controlled activity.

Recently, two multinational European Union funded projects have been completed. These are the Research Fund for Coal and Steel (RFCS) funded project "Real-Time Reconciliation and Optimization in large continuous mining Operations—RTRO-Coal" (Benndorf et al. 2019) and the European Framework Horizon 2020 funded project "Real-Time Mining" (Benndorf and Buxton 2019). The aim in these industry-driven projects has been to progress the technological development from a current Technology Readiness Level TRL3-5, which is that technology is tested and validated in a laboratory scale, to TRL6-7, where an integrated system is available and validated under industrial environment conditions in full scale.

1.2 Outline of the Book

This book is written to summarize key results and illustrate these through examples. The text is designed neither to cover all aspects of mineral source management nor to provide a complete description of the theory. It provides an overview and a description of concepts used for an implementation of the "Closed Loop Management in Mineral Resource Extraction" approach on an introductory level. Selected technical details, new methods, examples, and some practical aspects for operational implementation are discussed. With the referenced literature, the interested reader is provided with a starting point to dive deeper into the topic.

Chapter 2 introduces the closed-loop management concept of mineral resources focusing on grade control. Key constituents, including available sensor data, the resource and grade control model, and short-term mine planning approaches are described.

Chapter 3 presents a key constituent, which is the real-time resource or grade control updating engine. This allows integrating data from grade and production monitoring into grade control models, as soon as they become available. Based on the foundations of geostatistics, the theoretical background of applied data assimilation techniques is explained.

Chapter 4 presents three industrial-scale examples that demonstrate the capability of these techniques to continuously improve the prediction of ore quality attributes, mainly for the short- and medium-term periods.

Chapter 5 illustrates, how updated models can lead to better decisions in short-term mine planning and production control utilizing the concepts of simulation-based optimization. The application of the value of information concept to closed-loop mineral resource management is detailed. Results allow quantifying the benefits of a closed-loop approach.

References

J. Benndorf, C. Yüksel, M.S. Shishvan, H. Rosenberg, T. Thielemann, R. Mittmann, O.M. Lohsträter, M. Lindig, C. Minnecker, R. Donner, W. Naworyta, RTRO–coal: real-time resource-reconciliation and optimization for exploitation of coal deposits. Minerals **5**, 546-569. https://doi.org/10.3390/min5030509

J. Benndorf, M.W.N. Buxton, K. Nienhaus, L. Rattmann, A. Korre, A. Soares, A. deJong, N. Jeannee, P. Graham, D. Buttgereit, C. Gehlen, F. Eijkelkamp, H. Mischo, M. Sandtke, T. Wilsnack, Real-time mining-Moving towards continuous process management in mineral resource extraction. In *Proceedings of the 3rd International Future Mining Conference*, (AUSIMM, Sydney, 2015b), pp. 37–46

J. Benndorf, M. Buxton, A special issue on geomathematics for real-time mining. Math. Geosci. **51**, 845–847 (2019). https://doi.org/10.1007/s11004-019-09828-2

J. Benndorf, R. Donner, M. Lindig, O.M. Lohsträter, H. Rosenberg, S. Asmus, W. Naworyta, M.W.N. Buxton, Real-time Reconciliation and Optimization in large open pit coal mines (RTRO-Coal)–Final report. European Commission, Directorate-General for Research and Innovation, Directorate D-Industrial Technologies, 2019, p. 112. https://doi.org/10.2777/61704

J. Hou, K. Zhou, X.S. Zhang, X.D. Kang, H. Xie, A review of closed-loop reservoir management. Pet. Sci. **12**, 114–128 (2015). https://doi.org/10.1007/s12182-014-0005-6

J.D. Jansen, O.H. Bosgra, P.M.J. Van den Hof, Model-based control of multiphase flow in subsurface oil reservoirs. J. Process Control **18**, 846–855 (2008). https://doi.org/10.1016/j.jprocont.2008.06.011

J.D. Jansen, S.G. Douma, D.R. Brouwer, P.M.J. Van den Hof, O.H. Bosgra, A.W. Heemink, Closed-loop reservoir management. Paper SPE 119098 presented at the *SPE Reservoir Simulation Symposium*, (The Woodlands, USA, 2–4 February, 2009)

W.A. Shewhart, Economic control of quality of manufactured product. (ASQ Quality Press, 1931)

N. Zogovic, S. Dimitrijevic, S. Pantelic, D. Stosic, A framework for ICT support to sustainable mining -an integral approach. in *Conference: ICIST 2015 5th International Conference on Information Society and Technology*, (Kopaonik, Serbia, 2015), pp. 73–78. https://doi.org/10.13140/rg.2.1.2092.2723

Chapter 2
A Closed-Loop Approach for Mineral Resource Extraction

Abstract This Chapter introduces the closed-loop management concept of mineral resources focusing on grade control. This Chapter introduces first in the traditional mineral resource extraction chain and discusses some recent developments in production monitoring. Subsequently, it describes underlying models and optimization tasks in mining and introduces to the closed-loop mineral resource management (CLMRM).

2.1 Introduction

Mineral resource extraction can be seen as a discrete sequential physical extraction of small-scale blocks or smallest mineable units, which, depending on selectivity, can be on a meter by meter to tens-of-meters by tens-of-meter scale. The deposit and its properties remain static over time. Properties of the run-off-mine ore (ROM ore) stream are controlled by navigating the different extraction points or mining faces timely through the deposit. The aim is to extract blocks in a sequence and utilize production logistics such as blending piles or transport schedules in a way that production targets in terms of ore tonnage produced and related grades of value and deleterious elements are met. In the end, a good mine plan and production control impacts cash-flows and net present value (NPV). A focus of planning and production control is the reduction of in situ variability of ore in the deposit to a homogeneous product by means of blending at different stages and scales (e.g. Jupp et al. 2013; Benndorf 2013).

2.2 The Mineral Resource Extraction Process Chain

More often than not, the implemented approach in mine planning and operational control is a batch-type process of discontinuous and intermittent production monitoring and decision-making. Figure 2.1 shows the general steps in this linear value chain.

Fig. 2.1 Steps in a simplified mining value chain

Exploration consists of systematic data acquisition activities to define the spatial extent, geometry, geotechnical and hydrological conditions, and also the spatial distribution of value and deleterious elements within the deposit. Depending on the level of detail in exploration, different techniques are used including remote sensing, geological mapping, geochemical mapping, geophysics, and direct sampling by the means of drill coring. The latter one generates physical samples that are spatially constrained; the chemical or mineralogical composition is analyzed off-line in a laboratory. Although much effort is placed into different exploration activities, in general only a tiny part of the deposit is sampled. On average, the ratio between sampling volume and the volume of the deposit is 1:10.000.000. This number may deviate to be case-specific, however, the order of magnitude shows that our knowledge about the orebody after exploration is still rather limited and associated with a significant amount of uncertainty.

Based on exploration data, a 3D model of the orebody is generated, often using sophisticated geostatistical methods. The geometry of different geological zones of interest is modeled followed by the spatial distribution of value and deleterious elements within the zone. In fact, this does not only cover geochemical elements. In many operations also the spatial distribution of attributes relevant for processing and metallurgical extraction are modeled leading to so-called geometallurgical models (Domini et al. 2018). Examples are information related to the milling behavior, e.g., the BOND working index (Lynch et al. 2015). Section 2.4 will discuss in more detail the geostatistical modeling approach and also will differentiate between resource models and grade control models, which are used during operations management.

Considering modifying factors including mining, economic, marketing, legal, environmental, social, and governmental aspects, the resource is transformed into a reserve. Mineral reserves consist of recoverable reserves, meaning the part of the resource that can be technically and economically extracted considering above-mentioned factors. Planning activities typically include life-of-the-mine or long-term mine planning that defines the NPV of the project, which is further subdivided into medium-term and short-term plans that ensure feasibility in execution and aim to minimize cost. These plans define the extraction sequence of the areas and single SMUs within the deposit and the associated mine development effort. Short-term plans also consider concrete task schedules of resources, such as mining crews or mining equipment. Section 2.5 will discuss in more detailed mine planning approaches at different stages of the process.

Once a mine plan is generated, it will eventually be executed. The execution is guided by the short-term plan, operational activities are typically controlled on a shift-by-shift basis. One focus lies in grade control that comprises all activities that are concerned about grade streaming. The general aim is to decrease the grade variability, which is inherent in a deposit, to a level that is acceptable for costumers. Grade streaming involves several operational control decisions, including block classification, block extraction sequencing, stockpiling and blending. Each of these steps is based on the most recent information about the material characteristics of the extracted ore. In most operations, modern ICT infrastructure for production monitoring is implemented, such as localization sensors (GNSS) of extraction and transportation equipment or grade scanners above belt conveyors. Material tracking and monitoring systems allow for a detailed tracking back of ore to its source within the deposit.

The last step in the mining value chain is the shipping of ore to a customer or to the processing and beneficiation plant. Here, the actual consequence of attributes of the shipped material becomes obvious the first time. The behavior of the ore during processing and its actual composition often deviate from expectations. Deviations between model-based predictions and actual delivered ore can cause inefficiencies of downstream process steps or penalties with significant negative impact. Figure 2.2 shows an example of deviations of actual delivered calorific value (CV) content of coal from model-based predictions. It could be exemplary to any commodity, the frequency and magnitude of deviations are hard to predict.

Looking at this traditional process chain, it becomes apparent that it is a discontinuous batch-type process. In fact, the steps are performed by specialists in their field: exploration geologists, resource geostatisticians, mine planners, mining operations management, etc. Tasks in these different "silos" are mostly performed independently using expert software, communication between silos is limited and feedback is rare and may take long time spans of months or even years. Deviations from expectations are first seen in operations, which may be decades after data acquisition during exploration.

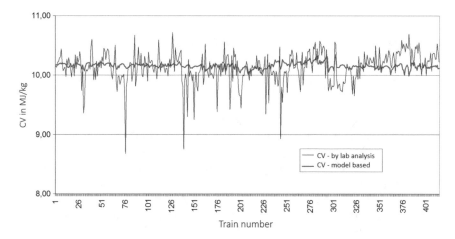

Fig. 2.2 Deviation of actual material attributes from model-based prediction on a train by train level

2.3 Production Monitoring—A Rich Data Base

With recent developments in production monitoring, online data capturing production performance provide an enormous source of information, alternative and in addition to exploration data. Literally, a flood of data is available, captured in a high spatial and timely resolution. Modern mining operations are equipped with sensors for online equipment locations and material tracking, online material characterization (texture, mineralogy, geochemistry, particle size distribution). Surveying and face mapping data are acquired regularly, equipment performance is monitored continuously (e.g., cutting energy, mill working index, metal recovery, etc.). These data are often soft, meaning of lower accuracy than direct sampling data from exploration, and only indirectly related to the underlying system models. However, due to their density in time and space they provide valuable information for comparing model-based expectations with reality.

To date, the sensor information is mainly utilized in feed-forward loops that are applied for downstream process control, such as supporting dispatch decisions, material sorting or blending on stockpiles. An immediate feedback of sensor information into the resource model and planning assumptions to continuously increase its certainty in prediction does not occur. In batch-type exercises, it is tried to improve the fit of mineral resource models to production monitoring observations and thus improve model assumptions. In mineral resource extraction the term reconciliation is used (e.g., Parker 2012). However, the ability to immediately feedback production monitoring data suggests a significant potential for improvement an operational efficiency. With increased certainty of prediction of ore qualities or material characteristics, the frequency of misclassification and unfavorable dispatch decisions is expected to decrease. Buxton and Benndorf (2013) quantify this value in the order of $5 Mio. per annum for an average-sized operation.

2.4 Resource and Grade Control Models

2.4.1 Resource Models

The steps of resource modeling aim to generate a 3D spatial model of deposit properties, such as vein thickness or grades. Resource models form the basis for resource and reserve evaluation and long-term mine planning and are thus one of the main decision-making tools along the mining value chain. Typically, these models are based on exploration drill hole data. The task in resource modeling is to utilize these to estimate deposit characteristics at non-sampled locations. Commonly, methods of geostatistics are used for this task. In geostatistics, the attribute under consideration is modeled through a random function. Sample data can be seen as a realization of the random variables at certain locations. Interpolation methods, such as Kriging (e.g., Goovaerts 1997), are used to estimate values and its uncertainty at not sampled locations, conditioned on available data. It is well recognized that estimation methods tend to smooth out local details of the spatial variation of the attribute. Typically, small values are overestimated whereas large values are underestimated (David 1977). Also, uncertainty about subsurface structures and parameters is not well captured. One of the major challenges in mining engineering is taking expensive design, planning, and operational decisions in the presence of very large uncertainties about the subsurface structure and the parameters that influence raw material production. To cope with this uncertainty, different possible scenarios about the subsurface models can be considered. In mineral resource extraction the terms "realizations" or "scenarios" are commonly used (e.g., Journel 1974; Dimitrakopoulos et al. 2002; Vann et al. 2012). Modern techniques of geostatistical simulation are used to generate these realizations or ensemble members. Typical requirements for simulation models are summarized in Table 2.1.

Given the aim of the model and data availability indicated in Table 2.1, solid mineral resource extraction uses data-driven approaches for local block prediction such as conditional simulation via turning bands or sequential Gaussian simulation (Journel 1974; Goovaerts 1997). The large extension of some mineral deposits requires computational efficient methods, such as direct block simulation methods (e.g., Marcotte 1994; Boucher and Dimitrakopoulos 2009; Desaisme et al. Deraisme et al. 2008) or the generalized sequential Gaussian simulation (Benndorf and Dimitrakopoulos 2018). The spatial correlation between grades is captured by methods using models of co-regionalization (e.g., Verly 1993; Soares 2001) or decorrelation methods (e.g., Rondon 2012; John and Donner 2018). Also, recently compositional approaches are being used for multivariate modeling, where the components represent the relative importance of the parts forming a whole (Tolosana-Delgado et al. 2019). The requirement for capturing in situ variability is in general sufficiently met using two-point statistics, such as variogram of spatial covariance function based methods.

Depending on the depositional environment and the subsequent diagenesis and structural deformation history (faulting, fracturing), it might be of importance to capture long-range connectivities or geometries. This requirement has led to the development of higher order (multipoint) geostatistical approaches. In particular

Table 2.1 Requirements for simulation methods to generate realizations and ensemble members

Requirements/Data	Criteria
Requirements on models	• Models need to capture in situ variability of geological zonation and grade distribution within zones. In general, multiple correlated grade attributes are of importance. The reproduction of connectivity is of lesser interest • Local focus: geological zonation and the local grade distribution in the different zones down to SMU scale
Data availability prior extraction	• Reasonable amount of information available enabling to prove continuity in geology and grade prior extraction • Data represent mostly direct samples (hard data) supported by geophysical interpretations

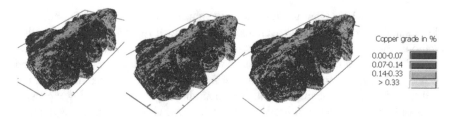

Fig. 2.3 Typical examples for realizations of a deposit (Figure from Benndorf and Dimitrakopoulos 2018)

training-image-based multipoint approaches have become very popular during the past decade; for overviews see, e.g., Caers (2011) or Mariethoz and Caers (2014). Examples are the SneSim (Strebelle 2002) and SGemS (Remy et al. 2009) algorithms, which are based on scanning a training image for multipoint configurations and reproducing similar images capturing the main geometrical features. Another popular technique is object-based simulation, based on the random generation of "geobodies", in the form of, e.g., elliptical "sand bars", sinusoidal "channels", or intersecting fault planes (Caers 2011, Pyrcz and Deutsch 2014).

Figure 2.3 shows an example of a simulated orebody, three realizations of a copper deposit.

2.4.2   Grade Control Models

While resource models are mainly used for decisions on a longer time scale and are modeled on a larger scale panel-support, short-term mine planning and production control is based on the so-called grade control models. These models are on the smallest mineable unit (SMU) scale, which is different from the larger panel support from the resource model. There are two challenges to be overcome.

The first one is the estimation of recoverable reserves prior extraction. Mine control decisions, such as block classification or material logistics, are based on the SMU scale. Prior to extracting a panel, no grade control data are available on the SMU scale. Therefore, the estimation of recoverable reserves has to be based on exploration data, which are rather scarce and do not have the level of spatial resolution required. If spatial interpolation methods are used to estimate SMU grades, the well-known smoothing effect introduces a bias and would result in a systematic estimation error leading to misclassifications. To correctly account for the change of support, methods such as uniform conditioning or localized uniform conditioning (Abzalov 2014) have been developed. Alternatively, recoverable reserves can be evaluated using geostatistical simulation based on a fine-scale grid with a subsequent re-blocking to the SMU size of interest (e.g. Peattie and Dimitrakopoulos 2013).

The second challenge is to integrate additional information obtained during production, as soon as these become available, such as received in dedicated grade control campaigns, face mapping, and sampling or blast hole sampling. Compared to exploration data, which often are of high precision, grade control data may be biased and of less precision. Geostatistical methods have been developed for integrating secondary data of different data sources. There is a large suite of methods available, e.g., several Co-Kriging or Co-Simulation methods (e.g., Goovaerts 1997). One particular method that has recently been applied for grade control modeling using data with different data qualities is kriging with variance of measurement error (Chiles and Delfiner 2009; Mariz et al. 2019).

2.5 Mine Planning Optimization

Based on the understanding of the resource and its associated uncertainty, different decisions or control options have to be taken during mineral resource extraction, ideally in an optimal way. These decisions and control options are different for different project stages and are thus based on a different amount of information available at the time. Some of the decisions involve large investments and can be hardly revised; some can be more flexible changed throughout the project. Table 2.2 provides examples of different time periods.

Figure 2.4 illustrates the robust optimization concept graphically for mining applications that considers geological uncertainty modeled using conditional simulation in geostatistics (e.g., Dimitrakopoulos 2011). The mine planning problem can generally be seen as a scheduling problem for discrete tasks applied to periods of extracting mining blocks or tasks. To solve this combinatorial optimization problem taking into account geological uncertainty, during the past decade various applications were published, including in large gold deposits (Godoy and Dimitrakopoulos 2004), meeting long-term production targets under joint-element uncertainty in iron ore (e.g., Menabde et al. 2018, Benndorf and Dimitrakopoulos 2013) or to define more robust ultimate pit limits (e.g., Vielma et al. 2009). Optimization problem sizes in mine

Table 2.2 Examples for control decisions to be optimized at different project stages

Time range	Decisions to be taken
Design (prior extraction)	• Ultimate pit limits (surface mining) or stope layout (underground mining) • Main infrastructure (e.g., shaft location and capacity) • Mining and processing capacity
Life of project to medium term	• Long-term extraction sequencing (e.g., pushback design)
Short range to production control	• Block classification • Short-term extraction sequence definition • Machine task scheduling • Dispatching, logistic, and stockpile management

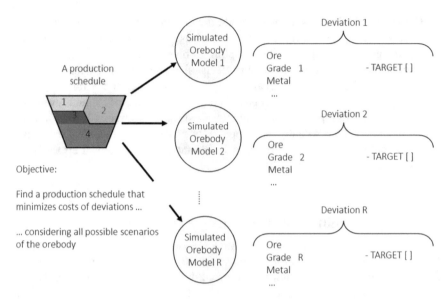

Fig. 2.4 Robust optimization in mining applications (reproduced after Benndorf and Dimitrakopoulos 2013)

planning are typically very large. For long-term scheduling problems, binary decision variables have to be defined per mining block and extraction period leading for average-sized problems to a number in the order of 10^6–10^7. Here, theoretical exact optimal approaches, such as integer or stochastic integer programming solutions, approach their computational limits. Recent work focuses on the development of optimization engines to solve extremely large combinatorial problems involving the complete value chain in mining complexes (e.g., Goodfellow and Dimitrakopoulos 2013). Combinatorial optimization techniques, such as simulated annealing (e.g., Kumral and Dowd 2005), tabu search (e.g., Lamghari and Dimitrakopoulos 2012) or evolutionary approaches (e.g., Gilani and Sattervand Gilani and Sattarvand 2016) provide computational efficient alternatives with a reasonable close-to-optimum result.

2.6 Making Use of Production Monitoring Data—Closing the Loop

While the previous discussions focus in general on robust optimized decision-making under geological uncertainty, this section will explore the use of additional data, available during production. This leads to closed-loop and real-time mining concepts.

Here "real-time" should be interpreted in the light of the relatively slow extraction processes. It refers to different time scales dependent on the control option to be taken. In case of dispatch decisions, this may in the order of minutes, for block classification tasks and production control tasks, decisions can be influenced by new data on a minute to hour scale. In general, production control options may be supported during a working shift, which is less than 8 h. In short-term planning application a shift-to shift basis seems a sensible time interval.

Figure 2.5 illustrates this concept for the Real-Time Mining approach, which is based on the plan–do–check–act (PDCA) iterative management cycle. It is general and applicable for surface mining and underground operations and can be interpreted as follows:

- *P—Plan and Predict*: Based on the mineral resource and grade control model, strategic long-term mine planning, short-term scheduling and production control decisions are made. Performance indicators such as expected ore tonnage extracted per day, expected ore quality attributes, and process efficiency are predicted.
- *D—Do*: The mine plan is executed.
- *C—Check:* Production monitoring systems continuously deliver data about process indicators using modern sensor technology. For example, the grade attributes of the ore extracted are monitored using cross-belt scanners. Differences between model-based predictions from the planning stage and actual measured sensor data are detected.
- *A—Act*: Differences between prediction and production monitoring are analyzed and root causes investigated. One root cause may be the uncertainty associated

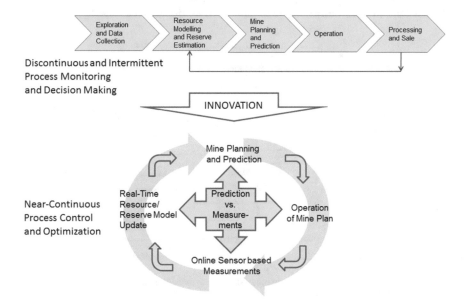

Fig. 2.5 The Real-Time Mining concept (after Benndorf et al. 2015b)

with the resource or grade control model to predict the expected performance. Another root cause may be the precision and accuracy of sensor measurements. Using innovative data assimilation methods, differences are then used to update the resource model and mine planning assumptions, such as ore losses and dilution. With the updated resource and planning model, decisions made in the planning stage may have to be reviewed and adjusted in order to maximize the process performance and meet production targets.

Figure 2.6 shows a system approach to closed-loop management for mineral resource management, which displays the key elements in the closed-loop system. This approach has been adapted from reservoir application, which demonstrated successful applications (Jansen et al. 2008, 2009).

The top of the figure represents the physical system consisting of the interaction between mining activities and the orebody. The center of the figure displays the system models. As discussed before, multiple models can be involved to quantify the large uncertainty in our knowledge of the subsurface.

On the right side of the figure, sensors that keep track of the processes and generate monitoring data from the system are displayed. These may be thought of as real sensors measuring production variables such as run-off-mine ore grades, production rates, and equipment performance. However, sensors may also be interpreted more abstractly as sources of information about the system variables, e.g., interpreted grade or hardness of the ore.

Optimization algorithms, indicated by a blue box and arrows, are shown on the left-hand side. Again, these may be interpreted as actual algorithms for production optimization influencing the job schedule of equipment, but also more abstractly as decisions in long-term mine planning, such as the pushback design. The state

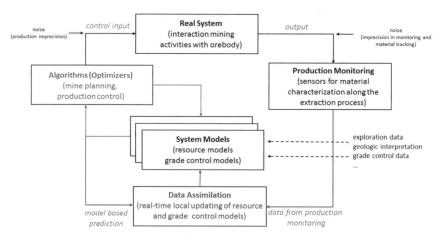

Fig. 2.6 Key elements of the closed-loop mineral resource management (adapted from after Jansen et al. 2008, 2009)

variables of the system, i.e., the grades at certain locations of the deposit, etc., are only known to a limited extent from the measured, usually noisy, output.

Finally, even the input to the system is only known to a limited extent; e.g., the exact digging line. The unknown inputs can also be interpreted as noise. Data assimilation can be used to reconcile the measured output with uncertain models to a certain extent. This is done by adapting the model parameters and model structure until the difference between measured and simulated data is minimized in some predefined sense, as indicated by the red box and arrows at the bottom. The two essential elements in the CLRM concept are, therefore, model-based optimization and decision-making (blue loop), and model updating through data assimilation (red loop). The latter one is the content of Chaps. 3 and 4, the optimization part of Chap. 5.

References

M.Z. Abzalov, Localized uniform conditioning (LUC): method and application case studies. J. South Afr. Inst. Min. Metall. **114**(3), 205–205 (2014)

J. Benndorf, R. Dimitrakopoulos, New efficient methods for conditional simulations of large orebodies. in *Advances in Applied Strategic Mine Planning*, (Springer, Cham, 2018), pp. 353–369

J. Benndorf, Investigating in situ variability and homogenisation of key quality parameters in continuous mining operations. Min. Technol. **122**(2), 78–85 (2013)

J. Benndorf, R. Dimitrakopoulos, Stochastic long-term production scheduling of iron ore deposits: integrating joint multi-element geological uncertainty. J. Min. Sci. **49**(1), 68–81 (2013)

J. Benndorf, M.W.N. Buxton, K. Nienhaus, L. Rattmann, A. Korre, A. Soares, A. deJong, N. Jeannee, P. Graham, D. Buttgereit, C. Gehlen, F. Eijkelkamp, H. Mischo, M. Sandtke, T Wilsnack, Real-time mining-Moving towards continuous process management in mineral resource extraction. in *proceedings of the 3rd international future mining conference*, (AUSIMM, Sydney, 2015b), pp. 37–46

A. Boucher, R. Dimitrakopoulos, Block simulation of multiple correlated variables. Math. Geosci. **41**(2), 215–237 (2009)

M.W.N. Buxton, J. Benndorf, The use of sensor derived data in optimization along the mine-value-chain. in *Proceedings of the 15th International ISM Congress*, (Aachen, Germany, 2013), pp. 324–336

J. Caers, *Modeling uncertainty in the earth sciences*, (Wiley, 2011)

J.P. Chiles, P. Delfiner, Geostatistics: *Modeling Spatial Uncertainty*, vol. 497, (Wiley, 2009)

M. David, *Geostatistical Ore Reserve Estimation*, (Elsevier, Amsterdam, 1977), p. 364

R. Tolosana-Delgado, U. Mueller, K.G. van den Boogaart, Geostatistics for compositional data: an overview. Math. Geosci. **51**(4), 485–526 (2019)

J Deraisme, J Rivoirard, P Carrasco Multivariate uniform conditioning and block simulations with discrete Gaussian model: application to Chuquicamata deposit. in *Proceedings of the Eighth International Geostatistics Congress*, eds. by J.M. Ortiz, X Emery (2008), pp. 69–78

R. Dimitrakopoulos, C.T. Farrelly, M. Godoy, Moving forward from traditional optimization: grade uncertainty and risk effects in open-pit design. Min. Technol. **111**(1), 82–88 (2002)

R. Dimitrakopoulos, Stochastic optimization for strategic mine planning: a decade of developments. J. Min. Sci. **47**(2), 138–150 (2011)

S.C. Dominy, L. O'Connor, A. Parbhakar-Fox, H.J. Glass, S. Purevgerel, Geometallurgy—a route to more resilient mine operations. Minerals **8**(12), 560 (2018)

S.O. Gilani, J. Sattarvand, Integrating geological uncertainty in long-term open pit mine production planning by ant colony optimization. Comput. Geosci. **87**, 31–40 (2016)

M. Godoy, R. Dimitrakopoulos, Managing risk and waste mining in long-term production scheduling of open-pit mines. SME Trans **316**(3) (2004)

R. Goodfellow, R. Dimitrakopoulos, Algorithmic integration of geological uncertainty in pushback designs for complex multiprocess open pit mines. Min. Technol. **122**(2), 67–77 (2013)

P. Goovaerts, Geostatistics for natural resources evaluation, (Oxford University Press on Demand, 1997)

J.D. Jansen, O.H. Bosgra, P.M.J. Van den Hof, Model-based control of multiphase flow in subsurface oil reservoirs. J. Process Control **18**, 846–855 (2008). https://doi.org/10.1016/j.jprocont.2008.06.011

J.D. Jansen, S.G. Douma, D.R. Brouwer, P.M.J. Van den Hof, O.H. Bosgra, A.W. Heemink, Closed-loop reservoir management. Paper SPE 119098 presented at the *SPE Reservoir Simulation Symposium*, (The Woodlands, USA, 2-4- February, 2009)

A. John, R. Donner, Integriertes Lagerstättenmanagement unter Anwendung geostatistischer Simulationsverfahren zur Modellierung sedimentärer Lagerstätten. in Stoffliche Nutzung von Braunkohle, (Springer Vieweg, Berlin, Heidelberg, 2018), pp. 117–130

A.G. Journel, Geostatistics for conditional simulation of ore bodies. Econ. Geol. **69**(5), 673–687 (1974)

Jupp, K., Howard, T.J., Everett, J.E. (2013) The role of precrusher stockpiling for grade control in mining, in Proceedings Iron Ore 2013, pp 203-214, (The Australasian Institute of Mining and Metallurgy: Melbourne)

M. Kumral, P.A. Dowd, A simulated annealing approach to mine production scheduling. J. Oper. Res. Soc. **56**(8), 922–930 (2005)

A. Lamghari, R. Dimitrakopoulos, A diversified Tabu search approach for the open-pit mine production scheduling problem with metal uncertainty. Eur. J. Oper. Res. **222**(3), 642–652 (2012)

A. Lynch, A. Mainza, S. Morell, Ore comminution measurement techniques, in *Comminution handbook. Spectrum series*, vol. 21, ed. by A. Lynch, (The Australian Institute of Mining and Metallurgy, Carlton Victoria, Australia, 2015), pp. 43–60

D. Marcotte, Direct conditional simulation of block grades, in *Geostatistics for the Next Century: Kluwer Academic Publishers*, ed. by R. Dimitrakopoulos (Dordrecht, The Netherlands, 1994), pp. 245–252

G. Mariethoz, J. Caers, Multiple-point geostatistics: stochastic modeling with training images (Wiley, 2014)

C.R.O. Mariz, A. Prior, J. Benndorf, Recoverable resource estimation mixing different quality of data. in *Mining goes Digital: Proceedings of the 39th International Symposium'Application of Computers and Operations Research in the Mineral Industry'(APCOM 2019)*, June 4-6, (CRC Press, Wroclaw, Poland, 2019), p. 235

M. Menabde, G. Froyland, P. Stone, G.A. Yeates, Mining schedule optimisation for conditionally simulated orebodies. in *Advances in applied strategic mine planning*, (Springer, Cham, 2018), pp. 91–100

H.M. Parker, Reconciliation principles for the mining industry. Min. Technol. **121**(3), 160–176 (2012)

R. Peattie, R. Dimitrakopoulos, Forecasting recoverable ore reserves and their uncertainty at Morila Gold Deposit, Mali: an efficient simulation approach and future grade control drilling. Math. Geosci. **45**(8), 1005–1020 (2013)

M.J Pyrcz, C.V. Deutsch, *Geostatistical reservoir modeling*, (Oxford university press, 2014)

N. Remy, A. Boucher, J. Wu, Applied geostatistics with SGeMS: a user's guide, (Cambridge University Press, 2009)

O. Rondon, Teaching aid: minimum/maximum autocorrelation factors for joint simulation of attributes. Math. Geosci. **44**(4), 469–504 (2012)

A. Soares, Direct sequential simulation and cosimulation. Math. Geol. **33**(8), 911–926 (2001)

S. Strebelle, Conditional simulation of complex geological structures using multi-point statistics. Math. Geol. **34**(1), 1–21 (2002)

J. Vann, S. Jackson, A. Bye, S. Coward, S. Moayer, G. Nicholas, R. Wolff, Scenario thinking: a powerful tool for strategic planning and evaluation of mining projects and operations. in *Project Evaluation 2012: Proceedings. Project Evaluation Conference*, May 2012, pp. 5–14

G.W. Verly, Sequential Gaussian cosimulation: a simulation method integrating several types of information. in *Geostatistics Troia'92*, (Springer, Dordrecht, 1993), pp. 543–554

J.P. Vielma, D. Espinoza, E. Moreno, Risk control in ultimate pits using conditional simulations. Proc. APCOM, 107–114 (2009)

Chapter 3
Data Assimilation for Resource Model Updating

Abstract One of the two core constituents of Closed-Loop Management for Mineral Resources is data assimilation for resource and grade control model updating. Similar to weather forecast models, the aim is to update the knowledge and forecast ability of the ROM ore as soon as new data from production monitoring become available. This chapter provides a formal description of the geostatistical foundations, a practical workflow and outlines one particular solution for updating. The theory is underpinned by three industrial-scale case studies and a discussion about practical aspects for operational implementation in Chap. 4.

3.1 The Data Assimilation Workflow

Figure 3.1 illustrates the general workflow for resource model updating, which consists of three sets of activities. The first set of activities relates to mine planning and forward modeling. The starting point is the in-situ mineral resource, which initially is not completely known to its extent, its geometry, and its inherent spatial distribution of grades and other attributes of interest. At different stages of the mine life, e.g., during exploration or grade control, samples are taken from the deposit. These form a database for resource or grade control modeling. As a result, spatially constrained estimated models are available, which typically discretize the deposit in blocks of equal geometry, e.g., on an SMU scale. These models represent our knowledge prior to extraction. In the terminology of updating these are called prior models, meaning that models are built based on information available prior to production. In a statistical view, the term prior model refers to the prior probability distribution of attributes within the deposit before some evidence from production monitoring is taken into account. The subsequent step of mine planning and production scheduling transforms the prior resource model into both, an estimation of mineral reserves and a model-based forecast of production expectations. Examples include expected ore tonnage produced and average grade of ROM ore per time unit.

The second set of activities relates to production monitoring. As discussed in the introduction, sensor networks deliver almost continuously information about

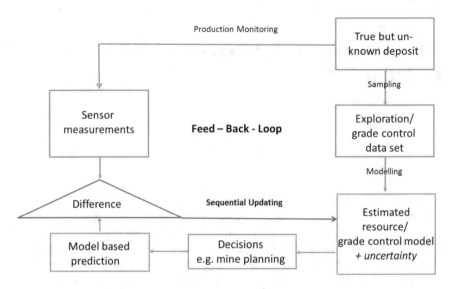

Fig. 3.1 Schematic process flow of resource model updating

the produced material. This information links directly or indirectly to material characteristics.

The third set of activities and the essential contribution of this chapter is the actual data assimilation step, the sequential model updating. This step uses both inputs, model-based prediction of production expectations and actual data values delivered from production monitoring. Differences shall be incorporated into the resource or grade control model for updating the prior model resulting in a posterior model. During the next production interval, this posterior model serves again as input for model-based prediction and is updated repeatedly in a sequential manner, as new data become available.

A Bayesian View

The updating described in this chapter can be seen as a particular implementation of the well-known Bayes Law, which is

$$P(H|D) = P(H) \cdot P(D|H) \big/ P(D) \tag{3.1}$$

where

- *P(H|D)* is the posterior probability. It describes the residual uncertainty in our resource or grade control model *H*, given the observed production monitoring Data *D*.
- *P(H)* is the prior probability. It describes the uncertainty of our resource or grade control model *H*, before any production monitoring data *D* are observed.

- *P(D|H)* is the likelihood. It describes the probability of the data *D*, given the prior uncertainty of the resource or grade control model. In other words, it describes how well the data *D* match with the prior model *H*.
- *P(D)* is the probability of observing the data *D*.

With this aside as background information, this chapter first provides a brief review of commonly used geostatistics methods for generating prior models. Next, the theoretical foundation of data assimilation for resource model updating is described in detail.

3.2 Geostatistical Concepts for Resource Modeling

The concepts presented here describe a typical workflow in resource or grade control modeling, from data to the model. Without claiming a complete review, major steps, model assumptions, and some selected methods are explained. For more detailed information, the interested reader is referred to typical textbooks (Goovaerts 1997; Chiles and Delfiner 2012; Deutsch and Rossi 2014; Abzalov 2016).

Resource modeling aims to generate a 2D or 3D spatial model of deposit properties, such as vein thickness, grades, or other attributes of interests. The task in resource modeling is to utilize exploration data to estimate deposit characteristics at non-sampled locations. Major steps involved in resource modeling are shown in Fig. 3.2 and include a statistical analysis of the data, the analysis and modeling of the spatial variability of attributes, known as variography, the spatial interpolation or simulation of the attribute at the non-sampled location and finally the model validation. The latter may lead to the conclusion that some fine-tuning in variogram modeling and parametrization of the estimation or simulation algorithm is necessary to obtain results that are more representative.

3.2.1 Explorative Data Analysis

3.2.1.1 Data Preparation

Resource models and thus also many planning related decisions in mining are based on data. If the database is defective, results will be defective too. Thus, having raw

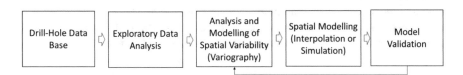

Fig. 3.2 Flowchart of resource modeling

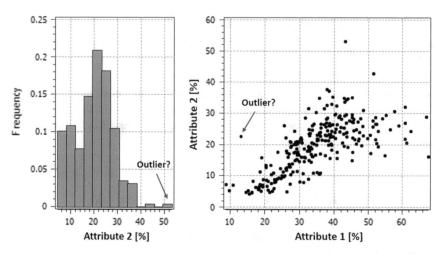

Fig. 3.3 Identification of outliers using histograms or scatterplots

data of exploration or grade control available, a first preparation step is to detect out-liers. To prevent the entry of defective data in the database, a corresponding QA/QC procedure should be in place. For exploration according to international standards, this is governed by international standards of resource and reserve reporting, such as the Canadian NI 43-101 or the Australian JORC Code. Sources of outliers include random errors due to precision of data sensing, systematic errors (bias), geologi-cal reasons (misinterpretation of domain or mineralization) or rough errors (due to mistakes). Random errors can be dealt with statistically or by using more precise analytics in monitoring. Systematic and rough errors should be identified and either corrected for (systematic errors) or eliminated (rough errors). To identify outliers, predefined tests can be constructed (e.g., the sum of certain elements has to be 100%), visual inspections of color-coded post-maps of exploration data can be performed or histograms and scatter plots may be used (Fig. 3.3). Often it is not clear, if an outlier is present or the large deviation is of random nature. Outliers can include valuable information about the characteristic of a deposit, such as the presence of local sulfur pockets or a gold nugget. Therefore, outliers should not be eliminated from the data set without good reason.

After the elimination of outliers in the database, the data support is to be standard-ized. This step refers to as compositing. The necessity of this step is because data may be obtained on different supports (sample length or volume), which imposes different statistical properties, in particular, the sample variance. Compositing is a sample homogenization by combining sample length to an approximately uniform support using a weighted average. After successful completion, data may be treated as one statistical population and statistical indicators are calculated.

3.2.1.2   Univariate Statistics

To obtain an overview of global data statistics, a common step is to calculate relative frequencies of data classes and plot these against its values to create a histogram. The shape of the histogram gives valuable insight into the behavior of the attribute (mean, dispersion, symmetry, outliers). An alternative plot is the cumulative frequency plot. The latter provides immediate information about global probabilities of being below a given threshold (Fig. 3.4).

For a quantitative description, statistical indicators are used. Statistical indicators of the central tendency of data include the following:

- Arithmetic mean m of the data:

$$m = \frac{1}{N} \sum_{i=1}^{n} z_i \tag{3.2}$$

where N is the number of data and z_i ($i = 1,...,N$) represent the single data values.
- Median M: the data value, which corresponds to a 0.5 probability of the cumulated frequency distribution.
- Mode D: data value of the highest frequency in the histogram.

Compared to the arithmetic mean, median and mode are robust indicators of central tendency in the presence of extreme values. Besides the central value, the spread of data is of interest. The spread gives an indication of the variability of attributes and is essential for a correct estimation of recoverable reserves. Statistical indicators describing the spread of the data include the following:

- Empirical variance of the data:

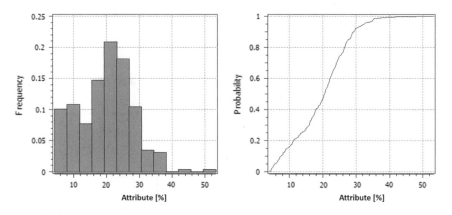

Fig. 3.4 Data histogram (left) and cumulative distribution (right)

$$\sigma^2 = \frac{1}{(N-1)} \sum_{i=1}^{N} (z_i - m)^2 \tag{3.3}$$

- The square root of the variance is the standard deviation σ. In the case of a normal distribution, approximately 68% of the data fall within an interval $m \pm \sigma$.
- Interquartile range:

$$\text{IQR} = Q_3 - Q_1 \tag{3.4}$$

The quartiles Q_1 and Q_3 represent the data values corresponding to 25 and 75% of the cumulated distribution. In the presence of extreme values, the IQR offers a more robust measure of spread compared to the variance or standard deviation.

3.2.1.3   Multivariate Statistics

If multiple attributes are of interest, the relation between them has to be considered. To investigate the relation between two and more sampled attributes, a multivariate data analysis is required. A correlation between attributes can be detected, such as between ash content and calorific value (CV) in Fig. 3.5. A typical graphical tool for the analysis between two attributes is the scatterplot. Samples, which contain an analysis value for both attributes under consideration, are plotted against each other. If the samples are very spread, there is no recognizable relation present. If samples

Fig. 3.5 Scatterplot showing the correlation between ash and CV

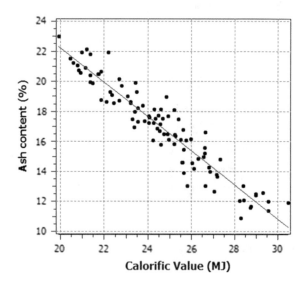

nearly follow a straight line as in Fig. 3.5, there exists a linear relationship, which, e.g., can later be used for improving the modeling result.

Following statistical indicators quantify the behavior between two attributes X and Y:

- *Covariance*:

$$C(X, Y) = \frac{1}{N} \sum_{i=1}^{N} (x_i - m_x)(y_i - m_y) \tag{3.5}$$

with the *arithmetic means*:

$$m_x = \frac{1}{N} \sum_{i=1}^{N} x_i \tag{3.6}$$

$$m_y = \frac{1}{N} \sum_{i=1}^{N} y_i \tag{3.7}$$

- *Coefficient of correlation*:

$$\rho = \frac{C(X, Y)}{\sigma_x \sigma_y} \tag{3.8}$$

with *standard deviations*:

$$\sigma_x = \sqrt{\frac{1}{(N-1)} \sum_{i=1}^{N} (x_i - m_x)^2} \tag{3.9}$$

$$\sigma_y = \sqrt{\frac{1}{(N-1)} \sum_{i=1}^{N} (y_i - m_y)^2} \tag{3.10}$$

A largely positive covariance or a coefficient of correlation near 1.0 indicates a strong positive correlation. A largely negative covariance or a coefficient of correlation near -1.0 indicates a strong negative correlation. Values close to zero indicate no correlation. Note that covariance and coefficient of correlation are solely measures of linear dependence. For nonlinear relation, other indicators such as the conditional expectation should be used.

3.2.2 Analyzing and Modeling Spatial Variability

Spatial variability of grades influences operating efficiency and required selectivity in resource extraction, metal recovery, ore losses, and dilutions or the predictability of ore characteristics in areas without drill holes. For a representative reserve estimation, it is essential being able to quantify the spatial variability. Figure 3.6 illustrates an example, where both Figs. (left and right) have identical univariate statistical indicators. Clearly, there is a difference in the spatial structure. The left figure is smooth, whereas the right one is highly variable. A quantified structural description of this behavior is the basis for a good spatial interpolation and for resource and grade control model updating.

The so-called *variogram* offers a descriptive tool for characterizing spatial variability. For pairs of exploration data spaced at different distances h, the variogram compares the values. Data, which are close to each other, show small differences. Data, which are spaced far from each other show larger differences. The *variogram* is calculated as half of the average quadratic distance between value $Z(x)$ at location x and $Z(x + h)$ at location $x + h$ for different distances vectors h.

$$\gamma(h) = \frac{1}{2N(h)} \sum_{i=1}^{N(h)} [Z(x) - Z(x + h)]^2 \qquad (3.11)$$

Note that x denotes the location vector with its components x, y, and z and $N(h)$ is the number of data pairs separated by distance h. In general, the distance between data can be described by distance vector h.

Plotting the calculated variogram values as a function of distance results in an experimental variogram. A typical example is shown in Fig. 3.7. The shape of the variogram can be described by the following parameters:

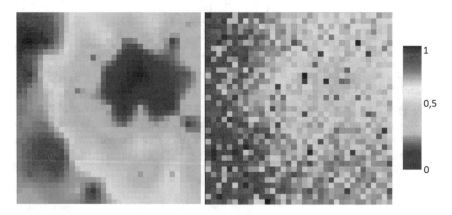

Fig. 3.6 Example for a spatially smooth and variable attribute

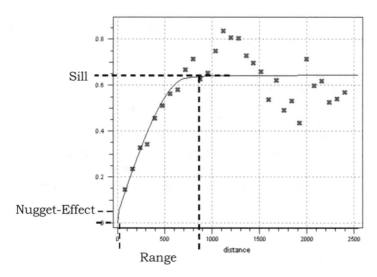

Fig. 3.7 Example for an experimental variogram

- *Nugget effect*: discontinuity at $h = 0$ m (in Fig. 3.7 approximately 0.05). This discontinuity is an expression of measurement uncertainty and infra-variability of the attribute. Two very close samples (h almost 0) can show the corresponding magnitude of differences.
- *Sill value*: max. plateau, which is reached by the variogram (in Fig. 3.7 approximately 0.65). For symmetric distributions the sill corresponds to the variance of the data. Two independently chosen data values (without considering distance) exhibit on average a corresponding squared difference.
- *Range*: defines the radius of influence of a sample (in Fig. 3.7 approximately 800 m). The range is a measure describing the possibility to transfer information from a sample to another location. Inside the range, two samples may be considered as statistically dependent or correlated. Outside of the range, there is no statistical dependence.

An alternative measure to the variogram is the spatial covariance. While the variogram characterizes the variability as a function of the distance h, the spatial covariance characterizes the spatial similarity as a function of the distance h.

$$C(h) = \frac{1}{N(h)} \sum_{i=1}^{N(h)} (Z(x) - m_x)(Z(x + h) - m_{x+h}) \tag{3.12}$$

Assuming second-order stationarity, the following relation applies to the spatial covariance and the experimental variogram. Here C(h = 0) corresponds to the covariance at h = 0.

$$\gamma(h) = C(h = 0) - C(h) \tag{3.13}$$

Some attributes may exhibit a different structural behavior in different directions. For a directional analysis, experimental variograms are calculated from data pairs, which are located in the specified direction. For a more detailed discussion the reader is referred to standard geostatistical textbooks (e.g., Journel and Huijgbregts 1978; Goovaerts 1997; Chiles and Delfiner 2012).

Having analyzed the experimental variogram, a theoretical variogram function has to be fitted (solid line in Fig. 3.7). The aim is to derive a functional description of the variogram values in dependence of separation distance h to be able to assess variogram values for all possible distances. This fitting is based on the experimental variogram. The later described interpolation and simulation methods require variogram values (or corresponding spatial covariance values) for any h, depending on the location to estimate and the data locations.

For variogram fitting, predefined functional models are available. The most commonly used include the Gaussian model, the spherical model, and the exponential model. These models are analytical expressions and include the parameters nugget effect C_0, sill value C, and range a, which have to be adjusted to fit the experimental variogram.

- Gaussian model:

$$\gamma(h) = C_0 + (C - C_0)\left(1 - e^{-\left(\frac{h}{3a}\right)^2}\right) \tag{3.14}$$

- Spherical model:

$$\gamma(h) = C_0 + (C - C_0)\left(1.5\frac{h}{a} - 0.5\frac{h^3}{a^3}\right) \; for \, h \leq a \; \text{and}$$

$$\gamma(h) = C \, for \, h > a \tag{3.15}$$

- Exponential model:

$$\gamma(h) = C_0 + (C - C_0)\left(1 - e^{-\left(\frac{h}{3a}\right)}\right) \tag{3.16}$$

The model-fitting exercise is the main step in spatial modeling and requires experience. The task is to fit a best model that captures all relevant structures and directional behavior. Multiple structures can be modeled by a nested variogram, meaning that multiple structures with separate parameters are added to a final model. The presence of a spatial trend requires a trend-elimination from data prior to the variogram analysis. The variogram is typically computed from stationary residual values.

3.2.3 The Random Function Model

Orebody forming processes, whether geological, geochemical, or physical, are often very complex and occur on different scales. An analytical description using functional relationships based on physical laws would require and a full understanding and an enormous set of parameters, which would be difficult to infer. An alternative is to use probabilistic or stochastic models. These types of models aim to mimic the spatial or spatiotemporal structure of the attribute under consideration without the necessity to fully understand all related geological or physical processes. Note that randomness in these models is not an expression of randomness in the orebody itself. Rather it is a measure of our ignorance regarding the complex related processes, which we only partly understand. In geostatistics, the attribute under consideration is modeled as a random process and described by a random function.

A spatial random function (RF) $Z(x)$ on domain D is a family of spatially correlated random variables $Z(x_1)$, $Z(x_2)$, ..., corresponding to all locations x_1, x_2, ... in the set D. These random variables are referred to as regionalized random variables and describe local uncertainty about the knowledge of an attribute within the deposit. If the domain consists of only a finite number of elements, then the random function $Z(x)$ is a finite family of random variables that can be specified by a finite-dimensional distribution (Yaglom 1987).

Random Variables and Random Functions

A random variable (RV) $Z(x_i)$ at location x_i ($i = 1, ..., N$) is a variable that can take a series of outcome values according to some probability distribution. Realizations of $Z(x_i)$ are denoted with lower case letters $z(x_i)$. Two types of random variables are distinguished: discrete and continuous random variables. Discrete random variables have a finite set of outcomes. They are completely specified by its probability function.

$$f(x_i; z_i) = Prob(Z(x_i) = z_i) \tag{3.17}$$

In the continuous case, the realization of a random variable can take values in a certain interval. Continuous random variables are completely specified by a probability density function $f(x_i; z_i)$ or its corresponding cumulative density function $F(x_i; z_i)$.

$$F(x_i; z_i) = P(Z(x_i) \le z_i) \tag{3.18}$$

The following discussions will focus on continuous RV's. The joint behavior between two RV's $Z(x_i)$ and $Z(x_j)$ at location x_i and x_j is defined by its joint probability density function or its corresponding cumulative density function $F(x_i, x_j; z_i, z_j)$.

$$F\big(x_i; z_i; x_j; z_j\big) = P(Z(x_i) \le z_i; Z\big(x_j\big) \le z_j) \tag{3.19}$$

A random function $Z(x)$ on a set D is completely defined by a finite-dimensional distribution $F(x_1,...,x_N; z_1,...,z_N)$ that is symmetric and compatible (Kolmogorov 1950; Yaglom 1987).

$$F(x_i, \ldots, x_N; z_1, \ldots, z_N) = P(Z(x_1) \le z_1; \ldots; Z(x_N) \le z_N) \qquad (3.20)$$

This function defines the spatial law of the N RV's at locations $x_1, \ldots, x_N$ and characterizes its joint uncertainty about the values $z_1, \ldots, z_N$. In geostatistics, the analysis of RF's is often limited to cumulative density functions that involve not more than two locations at a time. Only the first- and second-order moments of the spatial law are considered, which are expectation and covariance function or variogram:

$$E[Z(x_i)] = m$$

$$C(Z(x_i); Z(x_i + h)) = E[(Z(x_i) - E[Z(x_i)])(Z(x_i + h) - E[Z(x_i + h)])]$$

$$2\gamma(h) = VAR[Z(x_i) - Z(x_i + h)] \qquad (3.21)$$

$$\forall x_i \in D$$

For the inference of these parameters in spatial statistics, stationarity and ergodicity need to be assumed. A random function is said to be strictly stationary if its finite-dimensional distribution is invariant to translation vector $\mathbf{h}$ (Yaglom 1987).

$$F(x_i, \ldots, x_N; z_1, \ldots, z_N) = F(x_i + h, \ldots, x_N + h; z_1, \ldots, z_N) \qquad (3.22)$$

A random function is said to be weakly stationary or second-order stationary, if the moments of order one and two of the finite-dimensional distribution are invariant to translation $\mathbf{h}$, that is, the mean and variance are constant over the domain D and the covariance is a function only of the vector from x_i to x_j.

The assumption of ergodicity allows spatial sampling to be performed at one instant across a stationary zone with no change in the measured result.

3.2.4 Spatial Interpolation

Attributes of the deposit are known at few sample locations, e.g., at exploration drill hole locations. At unknown locations, e.g., mining blocks or future mining areas, the attribute of interest has to be estimated. Therefore, a regular grid is created covering the area of interest. Spatial interpolation allows for the estimation of values at each grid node. Figure 3.8 illustrates such a grid. The spacing of grid nodes should be chosen to meet requirements from the application of interest, e.g., to discretize the smallest mineable unit. On the other hand, also the data spacing is to be considered to define a reasonable spatial resolution.

To estimate the unknown values for all grid locations, e.g., for grid node P_0 in Fig. 3.9, the value at grid node P_0 is calculated as a weighted average or linear combination of known values at sample locations P_1 to P_4 in the local neighborhood.

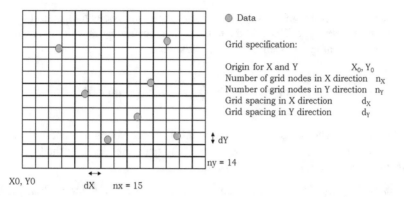

Fig. 3.8 Grid definition for spatial interpolation and simulation

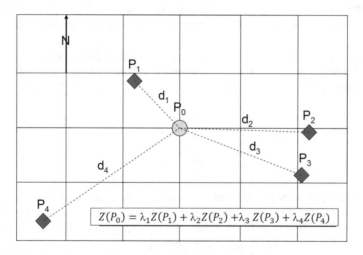

Fig. 3.9 Estimating values at grid nodes as weighted averages

The aim in spatial interpolation is to define the weights λ_i in such a way that the estimated value represents a so-called best linear unbiased estimator (BLUE). Unbiasedness ensures the absence of a global systematic error. On average, meaning over all grid nodes to estimate, the difference between estimated value $Z^*(x_i)$ and true but unknown value $Z(x_i)$ is 0. This is formulated in the unbiasedness condition.

$$E\left[Z^*(x_0) - Z(x_0)\right] = 0 \qquad (3.23)$$

The requirement "best" estimator leads to the condition of a minimum estimation variance σ_E^2. The average squared difference or variance between estimated value $Z^*(x_i)$ and true but unknown value $Z(x_i)$ has to be minimized.

$$\sigma_E^2 = VAR\left[Z^*(x_0) - Z(x_0)\right] \rightarrow MIN \qquad (3.24)$$

Geostatistics provides a set of linear estimators, which obey both conditions. The estimation of the value $Z^*(x_0)$ at the not sampled location x_0 as weighted average of the n adjacent sampled locations x_α is expressed as

$$Z^*(x_0) - m(x_0) = \sum_{\alpha=1}^{n} \lambda_\alpha [Z(x_\alpha) - m(x_\alpha)] \qquad (3.25)$$

Here, $m(x_0)$ and $m(x_\alpha)$ are the deterministic expectations or trend components of $Z^*(x_0)$ and $Z(x_\alpha)$. Different Kriging techniques are distinguished related to assumptions about this trend component:

- **Simple Kriging (SK)**: assumes the expectation or trend $m(x)$ of the random function as known and constant across the whole domain D considered: $m(x) = m$.
- **Ordinary Kriging (OK)**: accounts for local changes in expectation or trend $m(x)$ and limits its stationarity to the local neighborhood $W(x)$. The expectation or trend is unknown and is estimated from the values of the local neighborhood: $m(x) = m(x)$ in constant within $W(x)$.
- **Universal Kriging (UK)**: models the trend in a local neighborhood using a linear combination of elemental functional terms $f_k(x)$. The coefficients a_k are unknown and constant in the local neighborhood $W(x)$.

$$m(x) = \sum_{k=0}^{K} a_k(x) f_K(x) \qquad (3.26)$$

The detailed derivation of the Kriging system to solve at each grid nodes is skipped here. The interested reader is again referred to standard geostatistical text books (e.g., Journel and Huijbregts 1978; Goovaerts 1997; Chiles and Delfiner 2012). On the example of OK, the Kriging system is explained.

The mathematical derivation of the Kriging system accounts for the unbiasedness constraint. The Kriging estimator can be rewritten as

$$Z^*_{OK}(x_0) = \sum_{\alpha=1}^{n} \lambda_\alpha Z(x_\alpha) + \left[1 - \sum_{\alpha=1}^{n} \lambda_\alpha\right] m(x) \qquad (3.27)$$

If the sum of Ordinary Kriging weights equals 1

$$\Sigma_{\alpha=1}^{n} \lambda_\alpha = 1 \qquad (3.28)$$

the unbiasedness condition

$$[Z^*(x_0) - Z(x_0)] = E\left[\sum_{\alpha=1}^{n} \lambda_\alpha Z(x_\alpha) - m(x)\right] = \sum_{\alpha=1}^{n} \lambda_\alpha m(x) - m(x) = 0$$

$$(3.29)$$

is met. The OK estimator reduces to

$$Z_{OK}^*(\boldsymbol{x}_0) = \Sigma_{\alpha=1}^n \lambda_\alpha Z(\boldsymbol{x}_\alpha) \tag{3.30}$$

and does not require a prior definition of the mean $m(\boldsymbol{x})$. To determine optimal weights, the estimation variance (Eq. 3.24) has to be minimized. Substituting expression (Eq. 3.30) in expression (Eq. 3.24), an explicit mathematical expression as function of the Kriging weights can be derived. To introduce the unbiasedness constraint (Eq. 3.28) in this optimization task, a Lagrangian factor μ_{OK} is used, multiplied with the unbiasedness condition.

$$\sigma_E^2 = \Sigma_{\alpha=1}^n \Sigma_{\beta=1}^n \lambda_\alpha \lambda_\beta C(\boldsymbol{x}_\alpha; \boldsymbol{x}_\beta) + C(0) - 2\Sigma_{\alpha=1}^n \lambda_\alpha C(\boldsymbol{x}_\alpha; \boldsymbol{x}_0) + 2\mu_{OK}(\Sigma_{\alpha=1}^n \lambda_\alpha - 1) \tag{3.31}$$

Determining the partial derivative of (Eq. 3.31) with respect to the weights and the Lagrangian factor, and setting these to zero, results in the OK system. This ensures optimal Kriging weights under the unbiasedness condition and can be expressed by a multiplication of two matrices.

$$\lambda = \boldsymbol{C}^{-1} \boldsymbol{C}_0 \tag{3.32}$$

Here, λ represents the vector of optimal Kriging weights, $\boldsymbol{C}^{-1}$ is the inverse of the covariance matrix containing the spatial covariance values between all n data. Vector $\boldsymbol{C}_o$ contains the covariance values of spatial covariances between all n data and the location to estimate $\boldsymbol{x}_0$.

$$\boldsymbol{C} = \begin{bmatrix} C(\boldsymbol{x}_1;\boldsymbol{x}_1) & \dots & C(\boldsymbol{x}_1;\boldsymbol{x}_n) & 1 \\ \vdots & \ddots & \vdots & \vdots \\ C(\boldsymbol{x}_n;\boldsymbol{x}_1) & \dots & C(\boldsymbol{x}_n;\boldsymbol{x}_n) & 1 \\ 1 & \dots & 1 & 0 \end{bmatrix} \boldsymbol{C}_0 = \begin{bmatrix} C(\boldsymbol{x}_1;\boldsymbol{x}_0) \\ \vdots \\ C(\boldsymbol{x}_N;\boldsymbol{x}_0) \\ 1 \end{bmatrix} \lambda = \begin{bmatrix} \lambda_1 \\ \vdots \\ \lambda_n \\ \mu_{OK} \end{bmatrix} \tag{3.33}$$

Interpreting the OK system, three observations can be made:

- OK-weights are directly related to the spatial covariance between the data and the location to estimate $\boldsymbol{x}_0$. Data, which are closer to the point to estimate, obtain a larger weight as a function of the spatial structure modeled by the spatial covariance model.
- If some data locations are spatially close together, the covariance between these data is large. Due to the inverse of the corresponding covariance matrix $\boldsymbol{C}^{-1}$, these data will be reduced in weight. This mechanism takes care of the clustering effect that is the redundancy of information, when exploration data are very close to each other.
- The last row in the covariance matrix $\boldsymbol{C}$ contains a set of 1's, which forces the weights to sum up to 1. This condition results from the unbiasedness constraints and ensures an unbiased estimation.

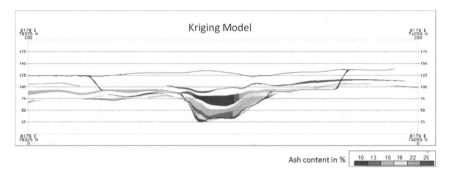

Fig. 3.10 Cross section of a spatially interpolated deposit model

If all grid nodes are estimated, an interpolated 2D or 3D model is available to describe the understanding of spatial grade distribution within a grade control panel or an orebody. Figure 3.10 shows a cross section through a coal deposit and the interpolated ash distribution.

Kriging also allows deriving a local measure of uncertainty associated with each grid node estimation. Based on previously discussed expression of estimation variance and the optimal OK-weights, the OK estimation variance can be expressed as

$$\sigma_E^2 = C(0) - \boldsymbol{C}_0^T \boldsymbol{C}^{-1} \boldsymbol{C}_0. \tag{3.34}$$

The Kriging variance provides a measure of estimation uncertainty, as discussed in the next section about conditional simulation in geostatistics, perhaps not the most realistic one.

3.2.5   Spatial Simulation

3.2.5.1   Introduction

The use of spatial interpolation methods tends to smooth out local details and does not allow for a realistic quantification of uncertainty in estimation and reproduction of in situ variability (e.g., Chiles and Delfiner 2012). For applications concerning uncertainties and risks as well as variability, an alternative is offered by using techniques of spatial conditional simulation. Figure 3.11 shows two realizations of a coal deposit generated by conditional simulation. Instead of generating one "best" model, these techniques generate an ensemble of multiple (10 or 50 or 100...) equally probable scenarios of the spatial distribution of attributes within the deposit. Each one could be the true deposit considering the information currently available. Each realization captures the spatial variability as analyzed with the variogram and does

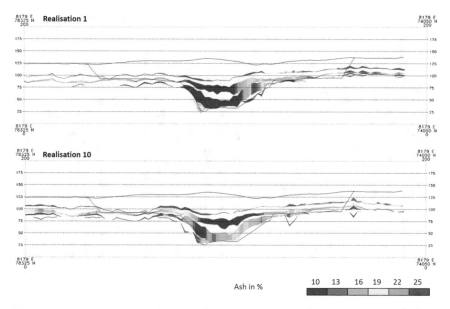

Fig. 3.11 Cross sections of two realizations of the simulated deposit model

not show smoothing effects. Local differences between scenarios provide a measure of uncertainty in prediction.

A short example shall illustrate the idea of using spatial simulations. The example is based on a study from Naworyta and Benndorf (2012). The task was to estimate recoverable lignite reserves in a polish mining field. The condition for mine-ability has been a CV of at least >8000 kJ/kg. For the part of the deposit to be evaluated, 20 simulated realizations and one estimated Kriging model have been generated for comparison. In addition, the actual production number is known for this part of the deposit. Figure 3.12 summarizes the results. Kriging overestimates recoverable reserves based on the 8000 kJ/kg threshold. Instead of 90%, as predicted by Kriging, in reality only 80% of the deposits met the requirement. This overestimation is a clear result of the inability to map spatial variability. Contrary, conditional simulation results are close to reality. The average value is very similar to the actual recovered reserves.

Methods of conditional simulation provide the means to model geological uncertainty as the basis for risk assessment and form the input for ensemble-based data assimilation method as described in Sect. 3.3.

3.2.5.2 Conditional Simulation of Gaussian Random Functions

There exists a vast amount of methods for conditional simulation. Acknowledging the fact that there are many methods available, each tailored to certain preferences, the subsequent section concentrates on a review of conditional simulation methods for

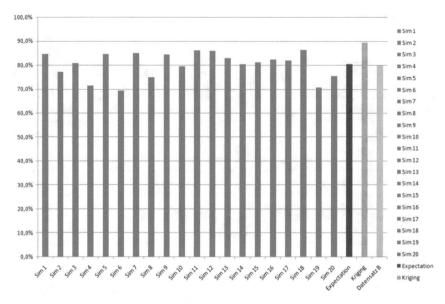

Fig. 3.12 Comparison of recoverable reserves estimated using Simulation and using Kriging with the true actual production (after Naworyta and Benndorf 2012)

Gaussian random the function. These methods are well studied and commonly used. The first two methods reviewed, conditional simulation via LU-decomposition (Davis 1987) and sequential Gaussian simulations (Isaaks 1990) are well-known methods, whereas the generalized sequential Gaussian simulation GSGS (Dimitrakopoulos and Luo 2004; Benndorf and Dimitrakopoulos 2018) is a generalization of both methods in order to increase the computational efficiency. Furthermore, the theoretical background of the direct block simulation (Godoy 2018), an extension of the GSGS, is reviewed. All these methods allow for the simulation of a single variable. Combining these methods with decorrelation approaches such as the minimum/maximum auto-correlation factor method or MAF (Debarats and Dimitrakopoulos 2000; Boucher and Dimitrakopoulos 2012) the above methods can be used for simulating vectors of correlated random functions at the point or block support.

Gaussian Anamorphosis

Methods discussed in this chapter are based on the assumption of normally distributed data, following a Gaussian distribution. Very often data do not follow a normal distribution. In these cases, original data have to be transformed into a normal distribution before applying simulation methods. Results from simulation methods have later to be back-transformed from normal space in the original data space. This pre- and post-processing step is known as Gaussian anamorphosis (e.g., Wackernagel 2013). Two commonly used approaches are the use of Hermite Polynomials and the so-called n-score transformation. The latter one is working with transformation tables (e.g.,

Goovaerts 1997). A more recently developed approach is the flow anamorphosis (van den Boogaart et al. 2017; Mueller et al. 2017).

Conditional Simulation via LU Decomposition

This well-known algorithm, introduced with data conditioning to geostatistics by Davis (1987), is based on the decomposition of the covariance matrix in an upper and lower triangular matrix using the technique of Cholesky decomposition. A symmetric and nonsingular, positive-definite covariance matrix C is partitioned as follows:

$$C = \begin{bmatrix} C_{11} & C_{12} \\ C_{21} & C_{22} \end{bmatrix} \tag{3.35}$$

where C_{11} is the covariance between data, C_{22} is the covariance between grid nodes to simulate and C_{21} or C_{12}^T, respectively, is the covariance between grid nodes and data.

Using Cholesky decomposition, the matrix can be decomposed in an upper and lower triangular matrix.

$$C = \begin{bmatrix} C_{11} & C_{12} \\ C_{21} & C_{22} \end{bmatrix} = LU = \begin{bmatrix} L_{11} & 0 \\ L_{21} & L_{22} \end{bmatrix} \times \begin{bmatrix} U_{11} & U_{12} \\ 0 & U_{22} \end{bmatrix} \tag{3.36}$$

Since C is a symmetric matrix, $L^T = U$. To generate an unconditional simulation, the lower triangular matrix has to be multiplied with a random vector of white noise $w \sim N(0,1)$.

$$\begin{bmatrix} y_1 \\ y_2 \end{bmatrix} = \begin{bmatrix} L_{11} & 0 \\ L_{21} & L_{22} \end{bmatrix} \times \begin{bmatrix} w_1 \\ w_2 \end{bmatrix} \tag{3.37}$$

It can be shown that the generated vector reproduces the variogram associated with the random function (Davis 1987). To condition the simulation, the vector y_1 needs to be replaced by the data vector z. Equation (3.37) can be rewritten as

$$\begin{bmatrix} z \\ y_2 \end{bmatrix} = \begin{bmatrix} L_{11} & 0 \\ L_{21} & L_{22} \end{bmatrix} \times \begin{bmatrix} L_{11}^{-1}z \\ w_2 \end{bmatrix} \tag{3.38}$$

Thus,

$$y_2 = L_{21}L_{11}^{-1}z + L_{22}w_2 \tag{3.39}$$

generates a realization of the random function conditioned on sample data z. The first term at the right side of Eq. (3.39) is identical to the simple kriging mean estimate, the L_{22} term to the square root of the simple kriging variance estimate, which correspond in case of the Gaussian random function to the conditional mean and conditional standard deviation (Anderson 1984). Regenerating the term $L_{22}w_2$ using different random numbers can generate multiple realizations.

The method is computationally efficient, as shown in Dietrich (1993) and Dimitrakopoulos and Luo (2004), the asymptotic bound of the runtime behavior is $O(N^3)$ and can only be marginally improved. A major disadvantage is, however, the limitation of the grid size to some few thousand nodes due to its extensive use of memory during the inversion of the covariance matrix.

Conditional Simulation via Sequential Gaussian Simulation

The method of sequential conditional simulation is based on the decomposition of the multivariate probability density function of a stationary and ergodic random function $Z(x)$ into a product of univariate conditional distributions (Rosenblatt 1952).

The sequential Gaussian simulation or SGS is the application of the decomposition of the multivariate distribution to a Gaussian random function (Isaaks 1990). Let $d_n = \{d(x_\alpha), \alpha = 1, \ldots n\}$ denote the set of conditioning data. Furthermore, let Λ_i be a data set such that $\Lambda_0 = \{d_n\}$ and $\Lambda_i = \Lambda_{i-1} \cup \{Z(x_i)\}$ including the set of conditioning data and the previous grid value that has been simulated. Following this notation, the sequential Gaussian simulation on a discrete grid D_N is based on the sampling the N-variate distribution conditional to the data set Λ_0 with a density function equal to the product of N single-variate conditional probability density functions.

$$f(\mathbf{x}_1, \ldots, \mathbf{x}_N; Z_1, \ldots, Z_N) | \Lambda_0 = f(\mathbf{x}_1; Z_1 | \Lambda_0)$$
$$\cdot f(\mathbf{x}_2; Z_2 | \Lambda_1)$$
$$\cdots$$
$$\cdot f(\mathbf{x}_N; Z_N | \Lambda_{N-1}) \tag{3.40}$$

Parameter of the conditional univariate distributions are sequentially obtained by calculating the conditional mean $E\{Z(x_i) | \Lambda_{i-1}\}$ and the conditional variance $Var\{Z(x_i) | \Lambda_{i-1}\}$ using Simple Kriging (Anderson 1984) and are given by

$$E\{Z(\mathbf{x}_i) | \Lambda_{i-1}\} = m + \mathbf{C}_{i\Lambda i-1} \mathbf{C}^{-1}_{\Lambda i-1 \Lambda i-1} (Z_{\Lambda i-1} - \mathbf{m}_{\Lambda i-1}) \tag{3.41}$$

$$Var\{Z(\mathbf{x}_i) | \Lambda_{i-1}\} = C + \mathbf{C}_{i\Lambda i-1} \mathbf{C}^{-1}_{\Lambda i-1 \Lambda i-1} \mathbf{C}_{\Lambda i-1i} \tag{3.42}$$

where m denotes the prior mean of $Z(x_i)$, C is the prior variance of $Z(x_i)$, $\mathbf{m}_{\Lambda i-1}$ denotes the vector of prior means of the conditioning data, $\mathbf{C}^{-1}_{\Lambda i-1 \Lambda i-1}$ is the inverse matrix of prior covariances between the conditioning data, $\mathbf{z}_{\Lambda i-1}$ represents conditioning data set Λ_{i-1} and $\mathbf{C}_{i\Lambda i-1}$ is the covariance vector between $Z(x_i)$ and the conditioning data Λ_{i-1}.

The implementation of the algorithm is based on the screen effect approximation. Only a subset $\lambda_{i-1} \subset \Lambda_{i-1}$ containing data in the local neighborhood is used to obtain the conditional distribution. The effect of using this approximation is that the Kriging system to be solved at each node does not increase with the number of simulated nodes. This decreases storage requirements during computation and enables SGS to simulate large grids. The asymptotic bound of the runtime behavior of SGS implemented with screen effect approximation is $O(Nv_{max}^3)$, where N is the number of grid nodes and v_{max} denotes the maximum number of nodes in the local neighborhood (Dimitrakopoulos and Luo 2004).

Sampling the obtained conditional distribution using Eq. (3.43) generates a realization of $Z(x_i| \lambda_{i-1})$

$$Z(\mathbf{x}_i|\Lambda_{i-1}) \approx E\{Z(\mathbf{x}_i)|\lambda_{i-1}\} + \sqrt{Var\{Z(\mathbf{x}_i)|\lambda_{i-1}\} \cdot w} \qquad (3.43)$$

where w is a random number $\sim N(0,1)$. The algorithm can be summarized in the following steps:

- Define a random path visiting N nodes to be simulated.
- At each node generate a simulated value using Eq. (3.43).
- Add this value to the conditioning data set.
- Loop through 2 to 3 until all nodes are visited.

Conditional Simulation via *Generalized Sequential Gaussian Simulation*

Dimitrakopoulos and Luo (2004) suggest the generalization of SGS, termed generalized sequential Gaussian simulation or GSGS, to enhance computational efficiency (Benndorf and Dimitrakopoulos 2018). The generalization is founded upon the observation that adjacent nodes share a common neighborhood (Fig. 3.13), and therefore the GSGS simulates groups of clustered nodes simultaneously instead of node-by-node. The use of groups of nodes amounts to the decomposition of the multivariate probability density function of $Z(x)$ into groups of products of univariate conditional probability density functions.

The simulation starts with the partitioning of the simulation grid D_N into k groups of v_j, $j = 1, ..., k$, clustered nodes and define N_j as number of nodes in the first j

Fig. 3.13 Shared neighborhood of group nodes (reproduced after Dimitrakopoulos and Luo 2004)

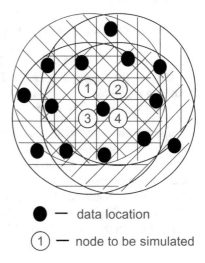

● — data location

① — node to be simulated

groups $N_j = \sum_{i=1}^{j} v_i, j = 1,..., k, N = N_k$. Then, the decomposition of the conditional probability density in equation into conditional densities for k groups becomes

$$f(\mathbf{x}_1, \ldots, \mathbf{x}_N; z_1, \ldots, z_N|\Lambda_0) \approx \prod_{i=1}^{N_1} f(\mathbf{x}_i; z_i|\lambda_{i-1}) \cdot \ldots \cdot$$

$$\prod_{i=N_k-1+1}^{N_k} f(\mathbf{x}_i; z_i|\lambda_{i-1}) \qquad (3.44)$$

where λ_{i-1} denotes the local conditioning data set including sample data and previously simulated nodes. The conditional mean vector and the posterior covariance matrix for group j are

$$E\left\{\mathbf{Z}(\mathbf{x}_i^{Nj}|\lambda_{i-1})\right\} = \mathbf{m}_j + \mathbf{C}_{j\lambda_{j-1}}\mathbf{C}_{\lambda_{j-1}\lambda_{j-1}}^{-1}\left(\mathbf{Z}_{\lambda_{j-1}} - \mathbf{m}_{\lambda_{j-1}}\right) \qquad (3.45)$$

and

$$\mathbf{Cov}\left\{\mathbf{Z}\left(\mathbf{x}_i^{Nj}|\lambda_{i-1}\right)\right\} = \mathbf{C}_{jj|\lambda_{j-1}} = \mathbf{C}_{jj} - \mathbf{C}_{j\lambda_{j-1}}\mathbf{C}_{\lambda_{j-1}\lambda_{j-1}}^{-1}\mathbf{C}_{\lambda_{j-1}j} \qquad (3.46)$$

where vector $\mathbf{m}_j$ denotes the prior means of group $\mathbf{Z}(\mathbf{x}_i^{Nj})$. $\mathbf{C}_{jj}$ denotes the prior covariance matrix of group $\mathbf{Z}(\mathbf{x}_i^{Nj})$, $\mathbf{m}_{\lambda j-1}$ represents vector of prior means of the conditioning data in λ_{j-1}, $\mathbf{C}_{\lambda j-1\lambda j-1}^{-1}$ is the inverse of the prior covariance matrix of the conditioning data λ_{j-1}, $\mathbf{Z}_{\lambda j-1}$ denotes the vector conditioning data in the local neighborhood λ_{j-1} and $\mathbf{C}_{j\lambda j-1}$ is the prior covariance matrix between $\mathbf{Z}(\mathbf{x}_i^{Nj})$ and λ_{j-1}.

The nodes of group j are generated using Cholesky to decompose the matrix $\mathbf{C}_{jj,\lambda j-1} = \mathbf{LL}^T$ and computed by following operation:

$$\mathbf{Z}(\mathbf{x}_i^{Nj}|\lambda_{i-1}) = \mathbf{m}_j + \mathbf{C}_{j\lambda_{j-1}}\mathbf{C}_{\lambda_{j-1}\lambda_{j-1}}^{-1}(\mathbf{Z}_{\lambda_{j-1}} - \mathbf{m}_{\lambda_{j-1}}) + \mathbf{Lw}_j \qquad (3.47)$$

where $\mathbf{w}_j$ is a vector white noise $\sim N(0,1)$. If the number of nodes in one group v is equal to one, the algorithm is identical to SGS. If the number of nodes in one group is equal to the whole grid size, the algorithm is identical to LU-decomposition. The implementation of the algorithm includes the following major steps:

- Define a random path visiting each group j of the grid and a sequential path visiting each node in a group.
- Define the local neighborhood of the current group.
- Calculate the conditional mean vector and conditional covariance matrix.
- Generate the simulated values of one group using Eq. (3.47).
- Add the simulated data values of the current group to the conditioning data set.
- Loop through steps 2 to 5 until all groups are simulated.

Benndorf and Dimitrakopoulos (2018) have shown that group sizes of 2×2 or 3×3 in two and $2 \times 2x2$ or $3 \times 3x3$ in three dimensions minimize the runtime behavior.

Direct Block Simulation

An extension of the GSGS algorithm is the direct block simulation (Godoy 2018). When simulating large grids in order of millions of nodes, values need to be retained as conditional information. This generates excessive memory requirements and leads to performance losses due to increased search times. To account for this drawback, a direct simulation method at block support scale has been developed. The method is based on GSGS, whereby the group of nodes discretizes a block.

Consider a normal score transformation of the random function $Y(\boldsymbol{x}_i)$ to $Z(\boldsymbol{x}_i)$. The regularized random function over a block support $Z_v(\boldsymbol{x}_j)$ with $\boldsymbol{x}_j \in R^3$, can be expressed as a linear average of $Z(\cdot)$ over the volume V, centered at the block center $\boldsymbol{x}_j$, and approximated by averaging the v internal nodes from a group:

$$Z_v(\mathbf{x_j}) = \frac{1}{V} \int_{x \in v} Z(\mathbf{u}) d\mathbf{u} \approx \frac{1}{v} \sum_{i=1}^{v} Z(\mathbf{x_i}) \tag{3.48}$$

Since the objective is to simulate block values $y_v(\boldsymbol{x}_j)$ in data space and not in Gaussian space $z_v(\boldsymbol{x}_j)$, after simulating a back-transformation from the Gaussian space into the data space needs to be performed. However, since the normal score transformation was performed using point values, there is no back-transformation for blocks of type $y_v(\mathbf{x_j}) = \Phi_v^{-1}(z_v(\mathbf{x_j}))$ available, unless restricting distribution assumptions are made. A solution to this problem is given by the approximation

$$y_v(\mathbf{x_j}) \approx \frac{1}{v} \sum_{i=1}^{v} \Phi^{-1}\left(z(\mathbf{x_i}|\lambda_{j-1})\right) \tag{3.49}$$

which is an averaging of all back-transformed internal nodes $y(\mathbf{x_i}|\lambda_{j-1})$ for $i = 1,\ldots,v$ of one group. To derive these values, the group $Z(\boldsymbol{x}_i^{Nj}) = (Z(\boldsymbol{x}_i), i = 1,\ldots,v)$ is first simulated. After simulating the internal nodes of a group and back-transforming these, the simulated block value is calculated as the average of the point values in Gaussian space and in data space, and, subsequently, point values are discarded. The simulated Gaussian block value is then added to the conditioning data set and the block value in data space is added to the results.

Considering the conditioning data, there are two types: point values Λ_i and block values included in the new subset Λ_i^V. With this definition and considering the screen effect approximation, the definitions of the GSGS formulation in Eq. (3.44) can be written in terms of point and block conditioning.

$$f(\mathbf{x_1}, \ldots, \mathbf{x_N}; z_1, \ldots z_N | \Lambda_0) \approx \prod_{i=1}^{N_1} f(\mathbf{x_i}; z_i | \lambda_{i-1}) \cdot \prod_{i=N_1+1}^{N_2} f(\mathbf{x_i}; z_i | \lambda_{i-1} \cup \lambda_1^V) \cdot \ldots$$

$$\cdot \prod_{i=N_{k-1}+1}^{N_k} f(\mathbf{x}_i; z_i | \lambda_{i-1} \cup \lambda_{k-1}^V) \tag{3.50}$$

To integrate the block support conditioning data, the algorithm is developed in terms of a joint-simulation. The second variable relates to the block value sequentially derived throughout the simulation process. The parameters of the successive conditional Gaussian distributions are obtained by solving a joint simulation system (Myers 1989), identical to joint point-block LU simulation. Let λ_{jV-1} be the joint neighborhood of points and blocks $\lambda_{j-1} \cup \lambda_{j-1}^V$, then the conditional mean vector and covariance matrix can be estimated from

$$E\left\{ Z\left(\mathbf{x}_i^{N_j} | \lambda_{jV-1}\right) \right\} = \mathbf{m}_j + \mathbf{C}_{j\lambda_{jV-1}} \mathbf{C}^{-1}_{\lambda_{jV-1}\lambda_{jV-1}} (\mathbf{Z}_{\lambda_{jV-1}} - \mathbf{m}_{\lambda_{jV-1}}) \tag{3.51}$$

$$\mathbf{Cov}\left\{ Z\left(\mathbf{x}_i^{N_j} | \lambda_{jV-1}\right) \right\} = \mathbf{C}_{jj|\lambda_{jV-1}} = \mathbf{C}_{jj} - \mathbf{C}_{j\lambda_{jV-1}} \mathbf{C}^{-1}_{\lambda_{jV-1}\lambda_{jV-1}} \mathbf{C}_{\lambda_{j-1}j} \tag{3.52}$$

where $\mathbf{m}_j$ and $\mathbf{m}_{jV-1}$ is the vector of prior means of group $Z(\mathbf{x}^j)$ and of the conditioning data, $\mathbf{Z}_{\lambda_{jV-1}}$ is the vector containing values of conditioning data λ_{iV-1}, including the original point data and previously simulated block values, $\mathbf{C}^{-1}_{\lambda_{jV-1}\lambda_{jV-1}}$ is the inverse of the prior covariance matrix of conditioning data $\mathbf{Z}_{\lambda_{jV-1}}$; $\mathbf{C}_{jj}$ denotes the prior covariance matrix of $Z(\mathbf{x}^j)$ and $\mathbf{C}_{j\lambda_{jV-1}}$ is the prior matrix of cross-covariances between $Z(\mathbf{x}^j)$ and $\mathbf{Z}_{\lambda_{jV-1}}$. The vector of simulated values is given by

$$Z\left(\mathbf{x}_i^{N_j} \middle| \lambda_{iV-1}\right) = \mathbf{m}_j + \mathbf{C}_{j\lambda_{iV-1}} \mathbf{C}^{-1}_{\lambda_{iV-1}\lambda_{iV-1}} (\mathbf{Z}_{\lambda_{iV-1}} - \mathbf{m}_{\lambda_{iV-1}}) + \mathbf{L}_{jl|\lambda_{iV-1}} \mathbf{w}_j \tag{3.53}$$

where $\mathbf{L}_{jl\,\lambda_{jV-1}}$ is the lower triangular matrix obtained by Cholesky decomposition of the $\mathbf{C}_{jjl\,\lambda_{jV-1}}$ of the simulated group j, $\mathbf{w}_j$ is a vector of white noise $\sim N(0,1)$ random numbers. The simulation of the internal nodes of each block is similar to GSGS. The only difference is the inclusion of conditioning data of different support scale, namely point values and block values. The implementation of the direct block simulation algorithm proceeds as follows:

- Define a random path visiting each of the blocks to be simulated.
- Normalize data.
- For each block, generate the simulated values in the Gaussian space of the internal nodes discretizing the block and back transform these into data space.
- Derive the simulated block value by averaging values of simulated nodes in one group in Gaussian space and in data space.
- Discard values of internal nodes and add the simulated block value in Gaussian space to the conditioning data set; keep the block value in data space as the result.
- Loop through steps 3 to 5 until all blocks are simulated.

A major practical advantage of the algorithm above is the decrease in memory allocation due to discarding internal points. Furthermore, the method takes advantage of the GSGS formalism and is thus a fast algorithm. Note that the method does not call for a block transformation function, which is often based on a global change-of-support model. Note also that the variogram validation at a block support scale is substantially more efficient than point support.

3.2.5.3 Conditional Simulation of Multiple Multiple Variables

In mining applications, often multiple attributes $Z(x) = [Z_1(x),...,Z_n(x)]$ of the ore-body need to be considered for project evaluation. For example, in iron ore projects, iron, but also phosphorus, alumina, silica, and loss of ignition (LOI) are essential for ore quality definition. These variables are generally cross-correlated, which needs to be taken into account during the simulation process. In general, there exist three approaches to simulate multiple correlated variables, which will be named and briefly explained.

Joint simulation: An intuitive adaption of the univariate case is the extension of the random function considered to a multivariate random function by the means of a joint probability distribution. The underlying Kriging system is extended and calls for the inclusion of all cross-covariances during the stochastic modeling process. This can be very demanding in runtime and storage requirements. Examples of this approach include the Vector Conditional Simulation (Meyers, 1989), the Joint Sequential Simulation of Multi-Gaussian Fields (Gomez-Hernadez and Journel 1993), or the Direct Sequential Co-simulation with Probability Distributions (Horta and Soares 2010).

Decorrelation: An alternative way of joint simulation of a random vector $Z(x)$ is the orthogonalization. The classic idea is to rotate the data $Z(x)$ to obtain uncorrelated factors $Y(x) = Z(x)A$, where A is the rotation matrix, and performing simulation independently for each factor. After, the simulated factors are rotated back into the space of the original data. The minimum/maximum autocorrelation factors (MAF) method, developed by Switzer and Green (1984) and later reviewed by Desbarats and Dimitrakopoulos (2000), is a double rotation approach that allows for orthogonalization of $Z(x)$ for the case, where modeling the sample matrix covariance is adequate up to two nested structures. The idea of MAF is to transform the multivariate observations $Z(x) = [Z_1(x),...,Z_n(x)]$ using a set of p orthogonal linear combinations

$$Y_i(\mathbf{x}) = \mathbf{a}_i^T \mathbf{Z}(\mathbf{x}), i = 1, \ldots, p \qquad (3.54)$$

Each transform $Y_i(x)$ exhibits a greater spatial correlation than the previously determined transforms $Y_j(x)$ under the constraint of orthogonality. The consideration of the spatial component of cross-correlations represents the major advantage of the MAF algorithm over conventional principal component analysis (PCA), which only de-correlates at lag zero and hence ignores the spatial correlation of data. Desbarats and Dimitrakopoulos (2000) have used MAF in geostatistics successfully for simulation of fifteen joint attributes. Further applications can be found in Boucher and Dimitrakopoulos (2004) at a point and block support.

The compositional approach: Compositional data quantitatively describe the relative weight, importance, or contribution of some parts of a whole. A relative importance of a component cannot be negative. The relative contributions may be captured by the constraint that all components sum up to a constant, e.g., 1. Compositional data analysis is defined as a set of statistical tools and geometrical concepts built with the purpose of allowing results to satisfy this constraint, also when only sub-compositions are analyzed. This imposes that results obtained by analyzing sub-compositions are not contradicting to those obtained analyzing the whole composition. In addition, scaling invariance is required. This property ensures that the information carried by a composition should be independent of the constant that scales it. The idea of compositional data analysis is to capture the relative information through a log-ratio transformation of the data. In the literature, several transformations can be found, such as the additive log-ratio (alr), the centered log-ratio (clr) or the isometric log-ratio (ilr). For detailed information, the reader is referred to standard literature such as from Pawlowsky-Glahn and Buccianti (2011). For an overview of compositional concepts applied to geostatistics, the reader is referred to Tolosana-Delgado et al. (2019).

3.3   Sequential Updating of Mineral Resource or Grade Control Models

3.3.1   Introduction

The utilization of available online production data for a rapid feedback and optimization of the process requires fast integration and processing of data, back-propagation of process information into the planning models and real-time decision support. Here, a framework for sequential feedback of sensor-derived online data into resource and grade control models is introduced. The approach is general and applicable to estimated and simulated resource and grade control models.

Several sequential updating algorithms have been proposed in the literature in the past, for example, Sequential Kriging or Co-Kriging (Vargas-Guzmán and Yeh 1999) to progressively update a spatial estimation model, when new information becomes available. This sequential estimator improves the previous estimate by using linearly weighted sums of differences between the new data set and previous estimates at sample locations. To update simulated realizations of spatial random fields, the method of Conditional Simulation of Successive Residuals—CSSR (Vargas-Guzmán and Dimitrakopoulos 2002; Jewbali and Dimitrakopoulos 2011) provides a solution.

This method updates the spatial realizations as new data become available. There are two limitations with this method: the location of future data and the sequence of data capturing have to be known and defined in advance and the covariance matrix has to be stored. This is inflexible and creates extensive computational costs when grids are large.

For model updating in a production environment based on sensor data, two additional challenges have to be overcome. First, sensor measurements from the whole process chain and the material characteristics under consideration may only be indirectly related. Second, the material stream passing a sensor station may be composed of different sub-streams originating from different extraction points in different parts of the deposit. Figure 3.14 illustrates this situation for two cases, a continuous open-pit mining system extracts ore from multiple benches, and an underground mine where ore is extracted and loaded to a conveyor belt system at different loading points. Common to both examples is that the material streams from several production areas are combined into one single stream of blended material leading to the bunker or the stock- and blending yard. In practice, sensors are often located above blended material streams, such as central conveyors. Therefore, the feedback algorithm has to be designed to track back the differences between model-based prediction and sensor measurement to the several production areas.

To solve this challenge, an adaption of a discrete Kalman-Filter approach is proposed. This is designed to reestimate the unknown state parameters of a system recursively on streams of noisy observation data. In the context of resource or grade control models, the state parameters may be interpreted as block grades or other block attributes of interest.

Applications of Kalman Filter approaches in a geoscientific context are found in many documented studies. To name just a few, in reservoir engineering, smart well and history matching concepts are designed using Kalman Filter approaches (Jansen et al. 2009; Heidari et al. 2011; Hu et al. 2012), in hydrogeology and geosciences different inverse problems are solved including the updating of permeability patters (Hendricks Franssen et al. 2011) or the data assimilation applied to an estuarine system (Bertino et al. 2002). The application of these techniques is not well established in solid mineral resource extraction applications yet, particularly not for updating resource and reserve models based on sensor data.

The following sections present the proposed formulation for the application of sequential updating of resource models in a mining environment. Subsequently, the approach is illustrated in a synthetic case study representing an exhaustively known environment and the performance will be assessed as a function of different mining system configurations and sensor measurement precisions.

3.3.2 A Framework for Sequential Resource Model Updating

In this section, the formal description of the proposed approach is given. The framework is designed to be applicable for both, estimated and simulated resource

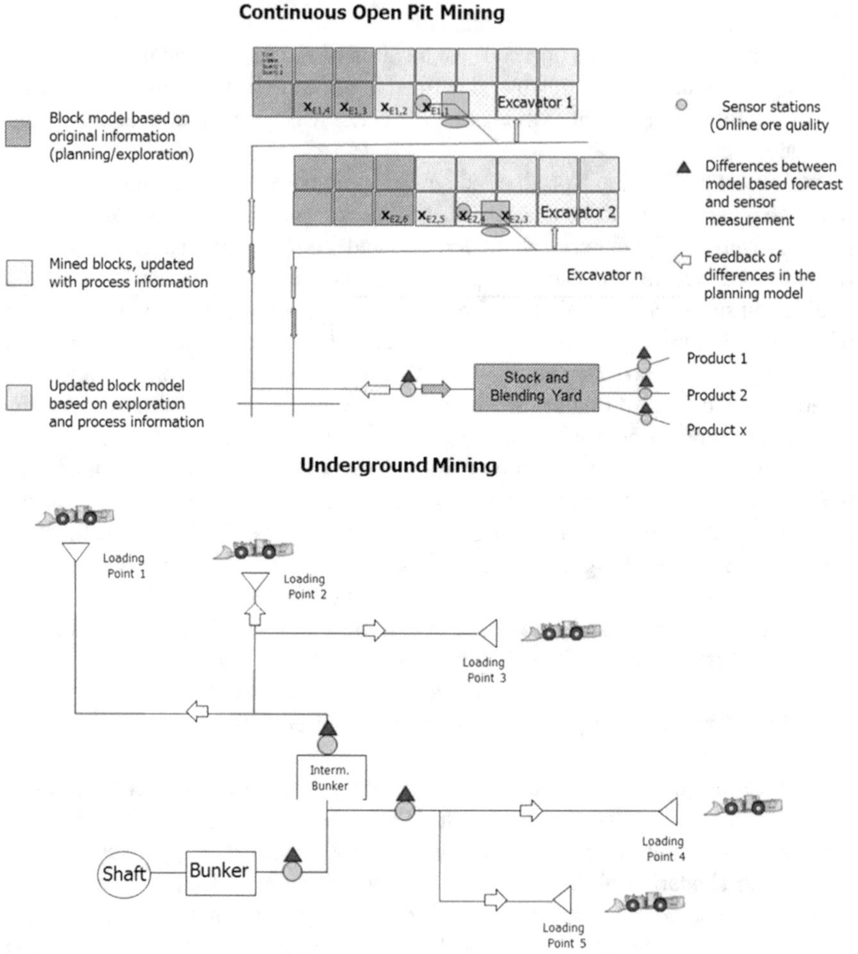

Fig. 3.14 Closed-loop concept in a continuous open-pit mining operation (top) and an underground operation (bottom)

and grade models. The first subsection provides the general derivation applicable to an estimated resource or grade control model according to Benndorf (2015). The second subsection provides an extension to the simulated resource or grade control models based on the ensemble Kalman Filter and mainly follows Wambeke and Benndorf (2017).

3.3.2.1   Derivation of a Mathematical Formulation for Sequential Resource Model Updating

Let vector $\mathbf{Z}(x)$ be a spatial random field with elements $Z(x_i)$ representing the material characteristic under consideration at locations x_i, with $i = 1,...,n$ being the index of discrete extraction locations or mining blocks. Further, let matrix $\mathbf{A}$ represent the production matrix describing the contribution of each of the mining blocks at x_i to the total production at a certain time interval t_j, with $j = 1,...,m$ that can be observed by a sensor.

$$\mathbf{A} = \begin{bmatrix} a_{1,1} & \cdots & a_{1,n} \\ \vdots & \ddots & \vdots \\ a_{m,1} & \cdots & a_{m,n} \end{bmatrix} \tag{3.55}$$

Elements $a_{i,j}$ can be interpreted as contributions of each mining block i to the produced material being on the conveyor belt, which will be eventually observed at some sensor station at time j. Matrix $\mathbf{A}$ is herein called production matrix and can be interpreted as an observation model, which links the block model $\mathbf{Z}(x)$ with sensor observations. In practice, this matrix can be obtained utilizing information from material tracking, e.g., satellite-based dispatch systems in open pit or a material flow model for stationary logistic elements and processing plants.

For illustration, the following numerical example is considered. In Fig. 3.14 (top), two excavators are exploiting the deposit. For five time intervals t_1 to t_5, the extraction rates of the excavator are recorded. Further, the location of the excavator is known using for a positioning system such as GNNS. Tracking systems are available that allow following the material stream through the process. In this way the recorded production rate can be unambiguously assigned to mining blocks. In total, there are eight mining blocks for each excavator in the next mining pass. For simplicity, consider a subset of four blocks per excavator with the material characteristics represented by the random variables $z(x_{E1,1})$ to $z(x_{E1,4})$ and $z(x_{E2,3})$ to $z(x_{E2,6})$. Matrix $\mathbf{P}$ summarizes the production of the two excavators during the five time intervals with respect to the mining blocks.

$$\mathbf{P} = \begin{bmatrix} & x_{E1,1} & x_{E1,2} & x_{E1,3} & x_{E1,4} & x_{E2,3} & x_{E2,4} & x_{E2,5} & x_{E2,6} \\ t_1 & 1,230 & 0 & 0 & 0 & 0 & 598 & 0 & 0 \\ t_2 & 983 & 115 & 0 & 0 & 0 & 572 & 0 & 0 \\ t_3 & 0 & 1,120 & 0 & 0 & 0 & 123 & 369 & 0 \\ t_4 & 0 & 1,093 & 0 & 0 & 0 & 0 & 489 & 0 \\ t_5 & 0 & 986 & 0 & 0 & 0 & 0 & 507 & 0 \end{bmatrix}$$

The observation model, captured in matrix $\mathbf{A}$, is computed standardizing each row by the total production per time interval t_i

$$A = \begin{array}{c|cccccccc} & X_{E1,1} & X_{E1,2} & X_{E1,3} & X_{E1,4} & X_{E2,3} & X_{E2,4} & X_{E2,5} & X_{E2,6} \\ \hline t_1 & 0.676 & 0 & 0 & 0 & 0 & 0.324 & 0 & 0 \\ t_2 & 0.589 & 0.069 & 0 & 0 & 0 & 0.343 & 0 & 0 \\ t_3 & 0 & 0.695 & 0 & 0 & 0 & 0.076 & 0.229 & 0 \\ t_4 & 0 & 0.691 & 0 & 0 & 0 & 0 & 0.309 & 0 \\ t_5 & 0 & 0.660 & 0 & 0 & 0 & 0 & 0.340 & 0 \end{array}$$

The multiplication of the production matrix A with an estimated grade control model, represented by vector $z^*(x)_t$, results in a model-based prediction z_t^* of characteristics of extracted material for time intervals t_i summarized in a time vector t

$$z_t^* = Az^*(x)_t \tag{3.56}$$

z_t^* is a vector of predicted values at a sensor location, computed by a linear combination of estimated block values $z^*(x_i)$. Note further that subscript t at $z^*(x)_t$ accounts for the information available up to time t and will be used to distinguish between a prior model and an updated posterior model, that will use the additional observations made during the observation period t, which includes observation subintervals t_j with $j = 1,...,m$.

Observations or measurements y_t of the characteristics of extracted material z_t are available for time intervals t_i. The relation to the true but unknown resource values $z(x)$ is generally described by

$$y_t = f(Az(x)) + v_t \tag{3.57}$$

where v_t is a random variable representing measurement noise. The vector v_t is assumed to be normally distributed with error covariance matrices $v_t \sim N(0, C_{v,v})$. The term $f(Az(x))$ expresses a general functional relationship between the sensor-based observation and the scanned material characteristic. If a linear observation model is assumed, as in most cases in mineral resource extraction, when blending is involved, Eq. (3.57) simplifies to

$$y_t = Az(x) + v_t \tag{3.58}$$

Kalman introduced a method describing a recursive solution to estimate the state of a stochastic process Z at time $t + 1$ based on the difference between a model-based prediction and available observations up to time t (Kalman 1960). In a similar way, for updating based on observations y_t, the new updated posterior resource model $z^*(x)_{t+1}$ is estimated as a linear combination of prior estimate $z^*(x)_t$ and the difference between an actual observations y_t during observation period t and the model-based prediction for the expected measurement result $Az^*(x)_t$

$$z^*(x)_{t+1} = z^*(x)_t + K(y_t - z_t^*) = z^*(x)_t + K(y_t - Az^*(x)_t) \tag{3.59}$$

The difference in Eq. (3.59), $y_t - Az^*(x)_t$, represents the gain in information about the in situ resource derived from sensor measurements and is called innovation. The matrix K is the so-called Kalman-gain matrix and can be interpreted as weighting factor that defines the influence of the detected differences to the estimation of the new posterior model $z^*(x)_{t+1}$. Since errors are associated with both, model-based prediction $Az^*(x)_t$ and measured observations y_t, the factor has to be defined accordingly. It has to be chosen to minimize the error in estimation of the new state $z^*(x)_{t+1}$ or the corresponding estimation variance.

Let the errors of the prior and the posterior resource model be defined as follows:

$$e(x)_t = z(x)_t - z^*(x)_t \tag{3.60}$$

$$e(x)_{t+1} = z(x)_{t+1} - z^*(x)_{t+1} \tag{3.61}$$

where $z(x)_t = z(x)_{t+1} = z(x)$ represent the true state. Since the deposit does not change over time, only the knowledge about the deposit, $z(x)$ remains static over time. The covariance matrices for state t and $t + 1$ can be obtained by

$$C_{t,t} = E\left[e(x)_t e(x)_t^T\right] \text{ and} \tag{3.62}$$

$$C_{t+1,t+1} = E\left[e(x)_{t+1} e(x)_{t+1}^T\right] \tag{3.63}$$

and capture the joint uncertainty in estimation about the components of $z(x)_t$ and $z(x)_{t+1}$, respectively, for mineral resource blocks before and after updating.

Combining (3.61) and (3.59) the estimation error can be expressed as

$$e(x)_{t+1} = z(x)_{t+1} - z^*(x)_{t+1} = z(x)_{t+1} - (z^*(x)_t + K(y_t - Az^*(x)_t)) \tag{3.64}$$

Substituting Eq. (3.64) into Eq. (3.63) results in the posterior variance–covariance matrix of the state vector $z^*(x)_{t+1}$

$$C_{t+1,t+1} = E\left[\left(z(x)_{t+1} - z^*(x)_{t+1}\right)\left(z(x)_{t+1} - z^*(x)_{t+1}\right)^T\right]$$
$$C_{t+1,t+1} = E\left[\left(z(x)_{t+1} - z^*(x)_t - K(y_t - Az^*(x)_t)\right)\left(z(x)_{t+1} - z^*(x)_t - K(y_t - Az^*(x)_t)\right)^T\right] \tag{3.65}$$

After a few simple reformulations, Eq. (3.65) can be expressed as

$$C_{t+1,t+1} = (I - KA)C_{t,t}(I - KA)^T + KC_{v,v}K^T \tag{3.66}$$

Equation (3.66) reflects the posterior error variance and has to be minimized with respect to K and can be rewritten as

$$C_{t+1,t+1} = C_{t,t} - KAC_{t,t} - C_{t,t}A^T K^T + K(AC_{t,t}A^T + C_{v,v})K^T \tag{3.67}$$

The partial derivative of the trace of $C_{t+1,t+1}$ with respect to K is

$$\frac{d(traceC_{t+1,t+1})}{dK} = -2(AC_{t,t})^T + 2K(AC_{t,t}A^T + C_{v,v}) \tag{3.68}$$

Setting the partial derivative to zero and solving it leads to the optimal Kalman-gain K.

$$K = C_{t,t}A^T(AC_{t,t}A^T + C_{v,v})^{-1} \tag{3.69}$$

An interpretation of Eq. (3.68) reveals the integrative character of the Kalman-gain. The second term is the inverse of two error sources, namely (a) the model prediction error, represented by the covariance matrix of the resource model $C_{t,t}$, which is propagated through the mining system by the production matrix A and (b) the measurement error, represented by the covariance matrix of the sensor-based measurement $C_{v,v}$. The first term again represents the model-based prediction. A comparison of potential magnitudes of the two error terms reveals that:

(1) if the prior model error is large and the measurement error is small, the Kalman-gain K tends to increase. The application to Eq. (3.68) shows that in this case a significant proportion of the difference between model-based prediction and sensor-based measurement is taken into account to update the resource model.

(2) if the prior model error is small and the measurement error large, the Kalman-gain K tends to decrease. The application to Eq. (3.68) indicates that the difference between model-based prediction and sensor-based measurement is then only slightly taken into account to update the resource model.

Integrating K in Eq. (3.68) with expression (3.67) results in

$$C_{t+1,t+1} = (I - KA)C_{t,t} \tag{3.70}$$

which expresses the posterior error variance as function of the Kalman-gain. Substituting (3.68) in Eq. (3.59) results in the update state vector

$$z^*(x)_{t+1} = z^*(x)_t + C_{t,t}A^T(AC_{t,t}A^T + C_{v,v})^{-1}(y_t - Az^*(x)_t) \tag{3.71}$$

Equation (3.69) and Eq. (3.70) describe the sequential updating of the resource model, as new sensor data become available. Figure 3.15 illustrates the complete sequential updating concept.

Note that the derived approach here is similar to the concept of sequential Kriging or Co-Kriging (Vargas-Guzmán and Yeh 1999) in geostatistics. The main extension is the ability to update the state vector $z(x)$ based on indirect observations, which represent linear combinations of attribute values $z(x_i)$ at certain locations i. This is achieved by the introduction of the observation model A, which links sensor observations and orebody attributes. As a result, the resource model can be updated based on

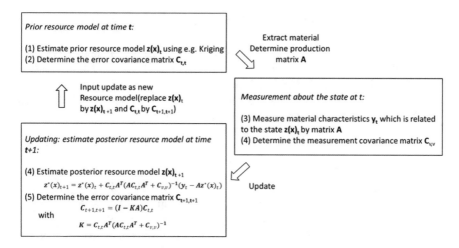

Fig. 3.15 Sequential updating process as new data become available

sensor observations, which detect a combined material stream coming from different sites in the mine and represent indirect information.

3.3.3 The Ensemble Kalman-Filter (EnKF) for Computationally Large and Nonlinear Relations

Due to the storage and propagation of the error covariance matrix, Kalman-Filter-based approaches suffer from computational efficiency, particularly when applied to large systems. In order to handle large problems with potential nonlinear dynamics, Evensen (1997) introduced the ensemble Kalman-Filter (EnKF) a method based on the Monte Carlo concept (Evensen 2003). Instead of propagating the covariance matrix in time using Eq. (3.69), the idea is to start with a finite set of $k = 1,...,K$ so-called ensemble members representing realizations $z_k(x)$ of the spatial random function $Z(x)$. Each of the k ensemble members is an equally probable representation of the spatial random field at time t. The whole ensemble describes the probability density. The initial set of ensemble members can be generated using techniques of conditional simulation in geostatistics, discussed in the first part of this Chapter.

Using Eq. (3.70), each ensemble members is propagated separately in time when new data y are available. The implementation of the EnKF includes the following computational steps:

- Forward operator A is applied to each ensemble member $z_k(x)$. Note that A can now be any complex operator that translates the regionalized random function describing the orebody into a measurable observation. Examples include mine simulators (Discrete Event Simulation) or material tracking systems (GNSS or underground positioning). Note that it is not anymore necessary to formulate the observation model A as a matrix.

- To maintain the variance and covariance structure of the ensemble members, the observations are treated as random variables (Burgers et al. 1998). For this reason, an error vector ϵ_k is added to the observation vector y used to update the several ensemble members k. The random error has a normal distribution with the mean of the first guess observation and a covariance equal to $C_{v,v}$

$$y_k = y + \varepsilon_k \tag{3.72}$$

- The calculation of the weighting factor K is performed empirically. Instead of storing the complete Covariance matrix $C_{t,t}$, only a finite set of ensemble members is kept. As described in Wambeke and Benndorf (2017), the following equations are used to approximate the necessary covariance-terms in Eq. (3.68) based on the i ensemble members

$$C_{t,t}A^T \cong \frac{1}{K}\sum_{k=1}^{K}(z_k(x) - \overline{z(x)}) \cdot (Az_k(x) - \overline{Az(x)})^T \tag{3.73}$$

$$AC_{t,t}A^T \cong \frac{1}{k}\Sigma_{k=1}^{K}(Az_k(x) - \overline{Az(x)})(Az_k(x) - \overline{Az(x)})^T \tag{3.74}$$

- The posterior covariance matrix can be approximated after solving Eq. (3.70) by

$$C_{t+1,t+1} \cong \frac{1}{K}\sum_{k=1}^{K}(z_k(x) - \overline{z(x)}) \cdot (z_k(x) - \overline{z(x)})^T \tag{3.75}$$

3.3.3.1 Dealing with Inbreeding

Inbreeding is essentially caused by the empirical estimation of covariances from Monte Carlo samples of limited size (Zhou et al. 2011). When occurring, the spread in the Monte Carlo sample becomes unrealistically small; in particular, the spread would be smaller than the difference between the sample mean and the unknown truth.

The problem generally arises when a group of grid nodes is being updated frequently during a certain period of time. The idea of configuring a pair of sequential updating cycles originates from Houtekamer and Mitchell (1998), who argued that the computation of the weights K and the evaluation of their quality (diagonal elements in $C_{t+1,t+1}$) would be based on exactly the same information when using a single Monte Carlo sample. As the assimilation cycle proceeds, the propagation of this dependent quality test might eventually result in a collapse of the Monte Carlo sample. To maintain sufficient spread, two connected updating cycles can be configured. The weights K computed from one cycle are used to assimilate data into the other cycle. In this way, each of the two cycles uses different Monte Carlo samples to estimate the weights and to evaluate the quality of an update. Implementation details about this so called double helix structure can be found in Wambeke and Benndorf (2017).

3.3.3.2 Dealing with Non-Gaussian Data

The updating approach presented so far is only optimal if all variables are multivariate Gaussian and if the forward observation model A is linear. For non-Gaussian cases, Wambeke and Benndorf (2017) presented a practical solution to implement the anamorphosis. Two constituents of the updating framework have to be transformed from the data space to the normal space, the prior model and the observations from production monitoring.

Model-specific anamorphosis function: To transform the model, commonly global normal score transformations are used (Goovaerts 1997; Deutsch and Journel 1992). However, due to the potentially developing non-stationarity, the underlying assumptions may be violated. This is, after a few updates, the global cumulative field distribution does not anymore represent the local updated conditions. To accommodate this situation, the approach suggested here transforms grid nodes or SMUs on an individual base according to node-specific Gaussian anamorphosis functions (Beal et al. 2010). These local functions are determined based on the local distribution, discretized by ensemble members, which are transformed into N(0,1) distributions.

Observation-specific anamorphosis function: To define an anamorphosis function for real observations, first from each ensemble member, model-based predictions $y_t^* = Az_t^*(x) + v_t$ are generated resulting in predicted observations. To account for measurement errors, the predicted observations are randomly perturbed, that is for each of the i predicted observations a random vector v_t is added, which is assumed to be normally distributed with error covariance matrices $v_t \sim N(0,C_{v,v})$. When applied to all ensemble members, a number of simulated predicted outcomes are available for each observation (perturbed predicted observations). These discretize a distribution of the observation, which can be transformed into Gaussian space applying an n-score transformation. Thus, an anamorphosis function has been defined that can be applied to real observation.

After the updating step, grid nodes are back-transformed using the inverse of the model-specific anamorphosis functions. A back-transformation of the observations is not required. Instead, updated predicted observations are obtained from a subsequent run of the forward predictor. Figure 3.16 summarizes the anamorphosis approach. A more detailed explanation is given in Wambeke and Benndorf (2017).

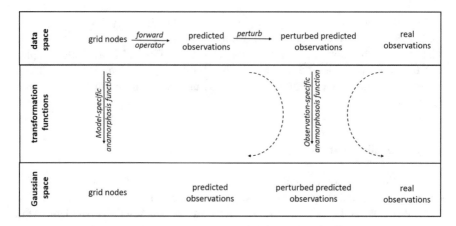

Fig. 3.16 Univariate transformation of grid nodes, perturbed predicted observations, predicted observations, and actual observations (reproduced after Wambeke and Benndorf 2017)

3.3.3.3   Discussion on Support Effects

Model and observations can be based on a different support. For example, grid node values of a model represent point data of grades and an observation from production monitoring may refer to an average grade over a blast block or an SMU. The presented EnKF approach, including the discussed Gaussian anamorphosis, accounts for these differences in support. This is because the underlying models are conditionally simulated models that capture in-situ variability. Any change of support can be modeled by relocking and is built in the updating engine by the observation function A that links the model and the observation. A more detailed discussion on support effects is provided in Prior et al. (2020a).

3.3.3.4   Note on the Extension to the Multivariate Case

An extension of the proposed approach to a multivariate case is not discussed here in detail. The utilization of joint-simulation approaches to generate the prior models appears a logical extension. Prior et al. (2020b) propose a compositional approach for data assimilation applied to resource models.

3.4   An Illustrative Example

3.4.1   Case Description

The following example aims to investigate the performance of the described updating methodology for different mining system configurations and precisions of sensor

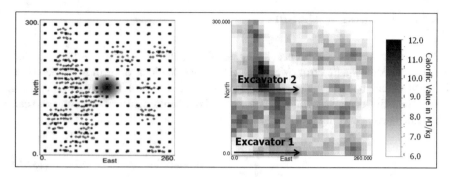

Fig. 3.17 Illustrative case study (left: database; right: OK block—model with indicated extraction sequence in case of two excavators)

measurements. It represents an artificial test case, which is built around the well-known and fully understood Walker Lake data set (Isaaks and Srivastava 1989). The data set (Fig. 3.17) is interpreted as a quality parameter of a coal deposit, for example, as calorific value. It is sampled irregularly at a spacing corresponding to an average of two reserve block-lengths. The blocks are defined with a dimension of 16 m x 16 m x 10 m. The block variogram is given with a spherical structure, range 50 m, nugget effect 0.4, and sill 0.6. Taking into account an assumed density of $2t/m^3$, one mining block represents a tonnage of 5,120t. Ordinary Kriging is used to generate a resource block model and the prior error covariance matrix, Generalized Sequential Gaussian Simulation is used to derive 100 realizations or ensemble members for the EnKF application. For simplicity, no dilution and losses are applied resulting in the reserve model being equal to the resource model. The resulting block model (Fig. 3.17) is used as the prior model.

Without loss of generality, the artificial block model is mined applying a continuous mining system equivalent to Fig. 3.14 (top), which consists of multiple bucketwheel excavators positioned at separate benches. Figure 3.17 shows the extraction sequence for the case of two excavators. Different digging rates are applied: excavator one extracts at a rate of 500t/h and excavator two at 1,000t/h. The material is discharged onto belt-conveyors positioned on the benches. These are combined into a single material stream at the central mass distribution point. The belt speed is assumed to be constant at 6 m/s.

The combined material stream is scanned by a sensor positioned above a central conveyor feeding the stock- and blending yard. Since no actual sensor data are available, virtual sensor data are generated. The artificial sensor data represent a 10-minute moving average corresponding to about 250t of production, and are composed of three components. Component one is the true block grade taken from the exhaustively known data set. Component two captures the volume variance relationship and corrects the smaller sensor measurement support of 250t to the mining block support of 5,120t by adding the corresponding dispersion variance (Journel and Huijbregts 1978). The third component mimics the precision of the sensor. For this case study, the relative sensor error varies between 1, 5, and 10%.

3.4.2 Evaluation Measure

The performance of the proposed sequential updating approach will be evaluated using the mean square error (MSE) related to the true block value of N blocks. Here the difference between estimated block value $z^*(x)$ and real block value $z(x)$ from the exhaustive data set is compared. The MSE is an empirical error measure and can be calculated according to

$$\text{MSE} = \frac{1}{N} \sum_{i=1}^{N} \left(z^*(\mathbf{x_i}) - z(\mathbf{x_i})\right)^2 \tag{3.76}$$

3.4.3 Results and Discussion

To evaluate the performance of different mining system configurations and different sensor precisions, the following cases are investigated:

(A) operating only one excavator using KF,
(B) operating two excavators simultaneously using KF and EnKF

Table 3.1 summarizes the parameters used in this experiment. In order to guarantee linear independency of rows in the production matrix $\mathbf{A}$, a cyclic component has been added to the extraction rates in the cases of two and three excavators. This cyclic behavior is typical for continuous mining equipment and can be observed in practice.

Figures 3.18 and 3.19 summarize the results of applying the Kalman-Filter to cases A and B. Figure 3.20 shows the results of the Ensemble Kalman-Filter applied to case B. The MSE is calculated separately for already mined blocks, blocks directly adjacent to the mined blocks and blocks which are two block-lengths away from mined blocks.

Figure 3.18 demonstrates the ability of the presented Kalman-Filter approach to decrease the uncertainty of predicting block values by updating based on sensor data. Considering the MSE, the following observations can be made:

- For mined blocks, the uncertainty almost vanishes. This is expected because in the case of one excavator the sensor measurements can be unambiguously traced back to the source block. Residual uncertainties remain due to sensor precision.

Table 3.1 Summary of parameters used for real-time updating of the resource model

	Extraction rate	Extraction mode	Sensor precision	Method
One excavator	E_1: 500t/h	Constant	1%, 5%, 10%	KF
Two excavators	E_1: 500t/h E_2:1000t/h	Cyclic	1%, 5%, 10%	KF and EnKF

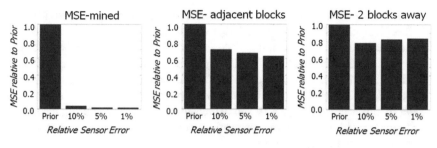

Fig. 3.18 Performance of the KF for updating the resource model in Case A

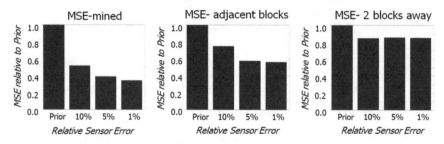

Fig. 3.19 Performance of the KF for updating the resource model in Case B

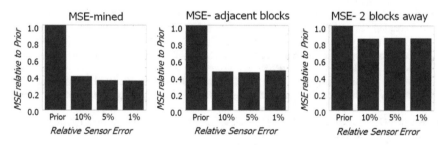

Fig. 3.20 Performance of the EnKF for updating the resource model in Case B

- Adjacent blocks are updated resulting in a significant improvement compared to the prior model. For high-precision sensors, this improvement leads to an approximately 40% decrease in the MSE. This improvement is due to the positive covariance between two adjacent blocks. In addition, the sensor clearly influences the result.
- Blocks in the second next row are also updated. Due to the larger distance and the corresponding smaller covariance, the effect is less obvious compared to directly adjacent blocks. It is, however, still significant.

Figure 3.19 shows the increased difficulty of the proposed method to track back the differences between sensor measurements and model-based predictions for combined material streams to the source blocks. The MSE's for mined blocks do not vanish completely; the remaining uncertainty can be interpreted as the limit of the filter for this specific application. It is expected, that with increased sensor sampling density, for example, every two or five minutes instead of ten minutes, the performance can be improved. Nevertheless, there is still a significant improvement in prediction for directly adjacent blocks and the next row of blocks compared to the prior estimate.

Figure 3.20 shows the example of the EnKF applied to case B. The results are very similar to those shown in Fig. 3.19 and demonstrate the validity of using the EnKF. Due to the limited problem size, observations concerning computational efficiency cannot be regarded as representative.

To summarize, results demonstrate a significant level of improvement in prediction by incorporating sensor data, in this case approximately 15%–40% relative compared to solely relying on exploration data. A more detailed study on the effects of blending ratios, sensor precision, and observation support can be found in Wambeke and Benndorf (2018).

References

M. Abzalov, *Applied Mining Geology*. Springer International Publishing (2016). https://doi.org/10.1007/978-3-319-39264-6

T.W. Anderson, *An Introduction to Multivariate Statistical Analysis*. (John Wiley & Sons, Inc., New York, 1984), 675 p

D. Beal, P. Brasseur, J.M. Brankart, Y. Ourmieres, J. Verron, Characterization of mixing errors in a coupled physical biogeochemical model of the north Atlantic: implications for nonlinear estimation using gaussian anamorphosis. Ocean Sci. **6**, 247–262 (2010)

J. Benndorf, Making use of online production data: sequential updating of mineral resource models. Math. Geosci. **47**(5), 547–563 (2015)

J. Benndorf, R. Dimitrakopoulos, New efficient methods for conditional simulations of large Orebodies. In: *Advances in Applied Strategic Mine Planning* (pp. 353–369). Springer, Cham (2018)

L. Bertino, G. Evensen, H. Wackernagel, Combining geostatistics and Kalman filtering for data assimilation in an estuarine system. Inverse Prob. **18**, 1–23 (2002)

A. Boucher, R. Dimitrakopoulos, Multivariate block-support simulation of the Yandi iron ore deposit Western Australia. Math. Geosci. **44**(4), 449–468 (2012)

G. Burgers, P. Jan van Leeuwen, G. Evensen, Analysis scheme in the ensemble Kalman filter. Mon. Weather Rev. **126**, 1719–1724 (1998)

J.P. Chiles, P. Delfiner, *Geostatistics, Modelling Spatial Uncertainty*, 2nd edn. (Wiley, New York, 2012)

M.D. Davis, Production of conditional simulations via the LU triangular decomposition of the covariance matrix. Math. Geol. **19**(2), 91–98 (1987)

R. Tolosana-Delgado, U. Mueller, K.G. van den Boogaart, Geostatistics for compositional data: an overview. Math. Geosci. **51**(4), 485–526 (2019)

A. Desbarats, R. Dimitrakopoulos, Geostatistical simulation of regionalized pore-size distributions using min/max autocorrelation factors. Math. Geol. **32**(8), 919–942 (2000)

C. Deutsch, A. Journel, *GSLIB Geostatistical Software Library and User's Guide* (Oxford University Press, Oxford, 1992)

C.R. Dietrich, Computationally efficient generation of Gaussian conditional simulation over regular sample grids. Math. Geol. **25**(1), 439–452 (1993)

R. Dimitrakopoulos, X. Luo, Generalised sequential Gaussian simulation on group size ν and screen—effect approximations for large field simulations. Math. Geol. **36**(5), 567–591 (2004)

G. Evensen, Advanced data assimilation for strongly nonlinear dynamics. Mon. Weather Rev. **125**, 1342–1354 (1997)

G. Evensen, The ensemble Kalman filter: theoretical formulation and practical implementation. Ocean Dyn. **53**, 343–367 (2003)

M. Godoy, A risk analysis based framework for strategic mine planning and design—method and application. In: *Advances in Applied Strategic Mine Planning* (pp. 75–90). Springer, Cham (2018)

J.J. Gómez-Hernández, A.G. Journel, Joint Sequential Simulation of Multigaussian Fields. In: *Geostatistics Troia'92*. (Springer, Dordrecht 1993), pp. 85–94

P. Goovaerts, *Geostatistics for Natural Resources Evaluation*. Applied Geostatistics Series. (Oxford University Press, New York 1997)

L. Heidari, V. Gervais, M. Le Ravalec, H. Wackernagel, History matching of reservoir models by ensemble Kalman filtering: The state of the art and a sensitivity study. Uncertainty Analysis and Reservoir Modeling: AAPG Memoir **96**, 249–264 (2011)

H.J. Hendricks Franssen, H.P. Kaiser, U. Kuhlmann, G. Bauser, F. Stauffer, R. Müller, W. Kinzelbach, Operational real-time modeling with ensemble Kalman filter of variably saturated subsurface flow including stream-aquifer interaction and parameter updating. Water Resour. Res. **47**(2) (2011)

A. Horta, A. Soares, Direct sequential co-simulation with joint probability distributions. Math. Geosci. **42**(3), 269–292 (2010)

P. Houtekamer, H. Mitchell, Data assimilation using an ensemble kalman filter technique. Mon. Weather Rev. **126**, 796–811 (1998)

L.Y. Hu, Y. Zhao, Y. Liu, C. Scheepens, A. Bouchard, Updating multipoint simulations using the ensemble Kalman filter. Comput. Geosci. **51**, 7–15 (2012)

E.H. Isaaks, The application of Monte Carlo methods to the analysis of spatially correlated data. Unpubl. PhD dissertation, Department of Applied Earth Sciences, Stanford University, Stanford, California, 213 p (1990)

E. Isaaks, R.M. Srivastava, *An Introduction to Applied Geostatistics*. (Oxford University Press 1989)

J.D. Jansen, S.D. Douma, D.R. Brouwer, van den P.M.J. Hof, O.H. Bosgra, A.W. Hemink, Closed-Loop Reservoir management. Paper 119098 presented at the SPE Reservoir Simulation Symposium. Woodlands (2009)

A. Jewbali, R. Dimitrakopoulos, Implementation of conditional simulation by successive residuals. Comput. Geosci. **37**, 129–142 (2011)

Joint Ore Reserves Committee (JORC). Australasian Code for Reporting of Exploration Results, Mineral Resources, and Ore Reserves (The JORC Code) (2012)

A.G. Journel, C.J. Huijbregts, *Mining Geostatistics* (Academic Press, London, 1978), p. 600

R.E. Kalman, A new approach to linear filtering and prediction problems. Trans. ASME J. Basic Eng. **82**, 35–45 (1960)

A.N. Kolmogorov, Foundations of the Theory of Probability: Chelsa. (New York, 1950), 71 p

D. E. Myers, Vector conditional simulation. In: *Geostatistics*. (Springer, Dordrecht 1989), pp. 283–293

U. Mueller, van den K. G. Boogaart, R. Tolosana-Delgado, A truly multivariate normal score transform based on lagrangian flow. In: Geostatistics Valencia 2016. (Springer, Cham 2017), pp. 107–118

W. Naworyta, J. Benndorf, Accuracy assessment of geostatistical modelling methods of mineral deposits for the purpose of their future exploitation—based on one lignite deposit. Mineral Resour. Manag. **28**(1), 77–101 (2012) (Polish)

NI 43–101. National Instrument 43-101, standards of Disclosure for Mineral Projects (NI 43-101). CIM SV 56—2013, National Instrument 43-101 Standards of Disclosure for Mineral Projects (2011)

V. Pawlowsky-Glahn, A. Buccianti, *Compositional Data Analysis*. (Wiley 2011)

A. Prior, J. Benndorf, U. Müller, Resource and grade control model updating for underground mining production settings. Math. Geosci. (2020a) (in print)

A. Prior, R. Tolosana-Delgado, van den K.G. Boogaart, J. Benndorf, Resource model updating for compositional geometallurgical variables. Math. Geosci. (2020b). (Accepted)

M. Rosenblatt, Remarks on multivariate transformation. Ann. Math. Stat. **23**, 470–472 (1952)

M.E. Rossi, C.V. Deutsch, *Mineral Resoure Estimation* (DOI, Springer, Dordrecht, 2014). https:// doi.org/10.1007/978-1-4020-5717-5

P. Switzer, A. Green, (1984). Min/max autocorrelation factors for multivariate spatial imagery. Dept. of Statistics, Stanford University, Tech. Rep. 6 (1984)

K.G. van den Boogaart, U. Mueller, R. Tolosana-Delgado, An affine equivariant multivariate normal score transform for compositional data. Math. Geosci. **49**(2), 231–251 (2017)

J.A. Vargas-Guzmán, T.C.J. Yeh, Sequential Kriging and co-kriging: two powerful geostatistical approaches. Stoch. Env. Res. Risk Assess. **13**, 416–435 (1999)

J.A. Vargas-Guzmán, R. Dimitrakopoulos, Conditional simulation of random fields by successive residuals. Math. Geol. **34**, 597–611 (2002)

H. Wackernagel, Multivariate geostatistics: an introduction with applications. (Springer Science & Business Media 2013)

T. Wambeke, J. Benndorf, A simulation-based geostatistical approach to real-time reconciliation of the grade control model. Math. Geosci. **49**(1), 1–37 (2017)

T. Wambeke, J. Benndorf, A study of the influence of measurement volume, blending ratios and sensor precision on real-time reconciliation of grade control models. Math. Geosci. **50**(7), 801–826 (2018)

A.M. Yaglom, Correlation theory of stationary and related random functions. (Springer, New York, 1987), 235 p

H. Zhou, J.J. Gomez-Hernandez, H.J. Hendricks Franssen, L. Li, An approach to handling Non-gaussianity of parameters and state variables in ensemble Kalman. Adv. Water Resour. **34**, 844–864 (2011)

Chapter 4
Updating Case Studies and Practical Insights

Abstract This chapter presents three industrial-scale applications of the updating framework introduced in Chap. 3. The first application in a gold mine applies ball mill monitoring data to continuously update the grade control model in terms of the Bond Work Index prediction. A more precise characterization of the input material in the comminution process allows for a more precise adjustment of control parameters. The second example demonstrates how coal quality data from a cross-belt online sensor are used in an open-pit mine to locally improve the coal quality model in real time. The third application focuses on the mineralogical characterization of a polymetallic ore vein in an underground mining operation. Data from hyperspectral imaging (Donner et al. 2019) and point sensor data using Laser-Induced Breakdown Spectroscopy (LIBS) or X-Ray Fluorescence (XRF) technologies (Desta and Buxton 2019) are used to continuously improve the prediction of vein geometry and the local mineral composition. All three case studies present industrial-scale results with a technology readiness level TRL 6 defined according to NASA's definition as the "System/sub-system model demonstration in an operational environment." This Chapter concludes with a summary of lessons learned and practical aspects when implementing a continuous feedback loop.

4.1 Use Case 1: Updating the Prediction of the BOND-Index for a Ball Mill

4.1.1 Case Description

Comminution is with up to 36% the largest energy consumer in the extraction process of raw materials (e.g., Ballantyne and Powell 2014). To separate the valuable mineral from the host rock, the mined ore is ground to the size of a few microns. Typical comminution equipment is a ball mill. An optimum adjustment of throughput performance requires an accurate prediction of the crushing behavior of the material, which is represented by a comminution index, such as the Bond Work Index (BWI). This index is a material parameter and describes the relationship between

© The Author(s), under exclusive license to Springer Nature Switzerland AG 2020
J. Benndorf, *Closed Loop Management in Mineral Resource Extraction*,
SpringerBriefs in Applied Sciences and Technology,
https://doi.org/10.1007/978-3-030-40900-5_4

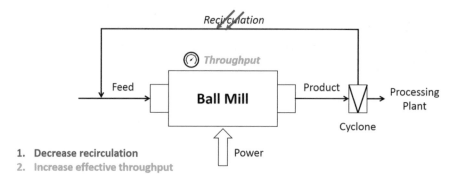

Fig. 4.1 Increasing efficiency in comminution by improved prediction of crushing behavior in a ball mill

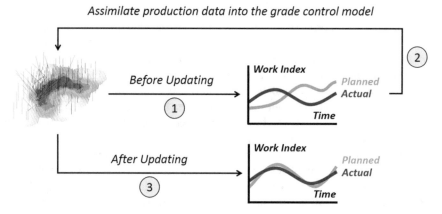

Fig. 4.2 Updating the Bond Work index in grade control models using online ball mill monitoring data

the hardness of the rock, throughput, size reduction of the particles, and the required crushing energy. The BWI defines the specific energy required to grind large rocks to a size of 100 μm (Lynch et al. 2015). If the actual value of this index deviates from the predicted or expected value, and if the process parameters are not optimally set, then losses and inefficiencies may be caused by increased recirculation of some of the material (Fig. 4.1).

To optimize mineral processing, specifically to control throughput through the ball mill, real-time updating of the grade control model is implemented in an opencast mine of Anglogold Ashanti in Western Australia (Wambeke et al. 2018). The aim is to improve the model-based prediction of the BWI by using data from the operational monitoring of the ball mill (Fig. 4.2). Input variables for model updating are the following:

- the grade control model for the BWI parameter,
- an observation or material-tracking model to link grade control block values with observations in the ball mill, and
- online ball mill monitoring data (energy consumption, throughput, grain size distributions).

4.1.2 Input Variables

Next, the individual ingredients for the model update are described in detail:

(a) *Short-term planning model (prior grade control model)*

The grade control model is used to spatially predict the BWI for each extraction block. These are defined for the benches considered with a dimension of 3 m × 3 m × 3 m. For each of these blocks, a corresponding material parameter is estimated from grade control samples. In general, laborious comminution tests in a laboratory are necessary to determine the work index from drilling samples. Such tests are cost-intensive and cannot be carried out for each production block. For this reason, comminution tests are performed for a representative set of training samples prior to extraction. At the same time, online sensors are utilized to determine the percentage contents of chemical elements of these samples. An applied principal component analysis provides essential variance-determining factors. Subsequently, a regression formula is developed that describes the relation between chemical components and the values of the comminution test. This allows for a preliminary estimation of the BWI from sensor-based geochemistry for each mining block. In order to generate the necessary spatial realizations or ensemble members of the prior grade control model, the Sequential Gaussian Simulation is applied.

(b) *Observation or material-tracking model*

To track the model values from the extraction blocks to the ball mill, it is necessary to integrate information from various implemented material-tracking systems. These are the following:

- Blast-block movements during blasting: using RFID tags, which position is scanned before and after the blast, the components of spatial movement and deformation of the mining block can be determined. On this basis, the spatial transformation of the information from the extraction block (grade control model) to the blasted heap is achieved.
- Mining equipment in open pits is often outfitted with satellite positioning (GNNS) for dispatch. This allows tracking loaded raw material and its properties through the logistic chain. The materials handling process within the pit allows to dump the material directly into the feed hopper of the processing plant (in Fig. 4.3—CRUSHER), or to temporarily store the material on stockpiles (in Fig. 4.3—ROM Finger). Therefore, temporary stockpile models are generated from material-tracking system information. This guarantees material tracking up to the processing plant.

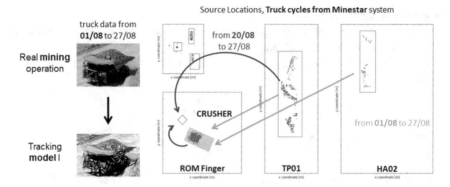

Fig. 4.3 Material tracking in the mining

- For material tracking through the processing plant up to the ball mill, a preliminary study has been conducted using Discrete Event Simulation (DES). Here, the entire process, the interaction of the individual elements as well as their random behavior is modeled (Fig. 4.4) to answer two questions: On average, how long does the material take from the primary crusher hopper to the ball mill? And, how precise is this time estimate? Results allow to conclude that considering intervals <4 h are not meaningful because for smaller time intervals the material tracking error would be too large to associate the material in the ball mill with the original mining blocks. For an interval of >4 h, this assignment error is of less significance.

(c) *Online data of the ball mill*

The installed sensors in the ball mill continuously measure the throughput, the energy consumption, as well as the grain size distribution of the input and the output material. From these parameters and using the relationship (Lynch et al. 2015)

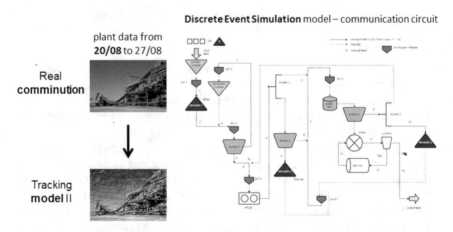

Fig. 4.4 Material tracking in the processing plant

$$\frac{P}{R} = W\left(\frac{10}{\sqrt{P_{80}}} - \frac{10}{\sqrt{F_{80}}}\right)$$

the Bond Work Index W can be estimated. P denotes the energy output [in kW], R is the flow rate [in t/h], F_{80} is the 80% quantile of the grain size distribution of the input material, and P_{80} is the corresponding quantile of the grain size distribution of the starting material [in µm].

4.1.3   Grade Control Model Updating

The updating of the grade control model is illustrated in Fig. 4.5. Based on grade control data, which are obtained by analyzing the chemical contents of the material from the blast holes, a prior grade control model is generated with the spatial distribution of the BWI. Using material tracking data from the pit and the processing plant, which serve as the forward operator A, a model-based prediction of material properties at the ball mill is made. Predicted forecasts are compared with the actual data from the operational monitoring of the ball mill. The application of the real-time updating formalisms provides the posterior grade control model updated with operational monitoring data. Considering the subsequent updating cycle, the previously generated posterior grade control model serves now as input, which is the prior model. In this way, a continuous updating process is achieved; as new data become available, the model is updated.

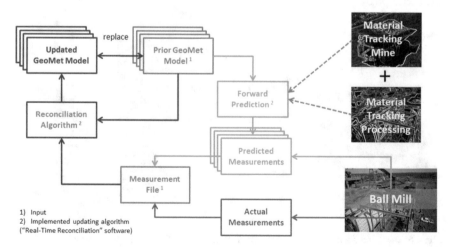

Fig. 4.5   Schematic representation of the process and data flow for updating the grade control model

4.1.4 Results

Figure 4.6 shows the prior and the posterior grade control models after 30 updates for the areas of two open-pit benches. The integration of online ball mill data visibly leads to a significant increase in the level of detail in spatial prediction. Figure 4.6 also demonstrates that in the posterior model, hard material can be better distinguished from soft material compared to the smoothed prior model, which underestimates the occurrence of high or low BWI values.

A case study has been conducted over several weeks. Figure 4.7 shows the model-based BWI prediction of the input material into the ball mill for different steps. The top figure corresponds to a time before the first update and thus represents the prior prediction. The middle figure represents 18 updates. The figure below shows the prediction based on the fully updated model. In addition to the predictions and their uncertainties, which are shown as a Box–Whiskers plot for each time interval, diagrams also contain the actual observation value of the operational monitoring, based on the 4-hour intervals. The bars under the diagram illustrate the model improvements compared to the prior model for the considered 4-h batches. Clearly, the update significantly improves the prediction of the blocks mined in the past. Only a small residual uncertainty remains. Also, the prediction of future blocks to be mined improves significantly. Over the next two days, the improvement is about 25–40% compared to the not-updated model. As a secondary result, a spatial model of the BWI, for blocks that have already been reconciled with the operating process, is quickly available. This offers the opportunity to evaluate this reconciled model against the resource model and builds up an operational database to learn from short-term to improve long-term predictions.

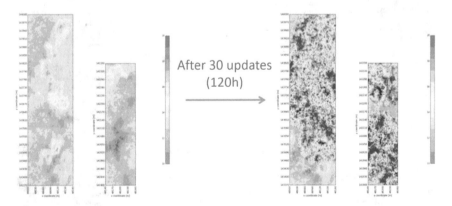

After 30 updates
(120h)

Fig. 4.6 Prior and posterior model of the Bond Work Index of two open-pit benches

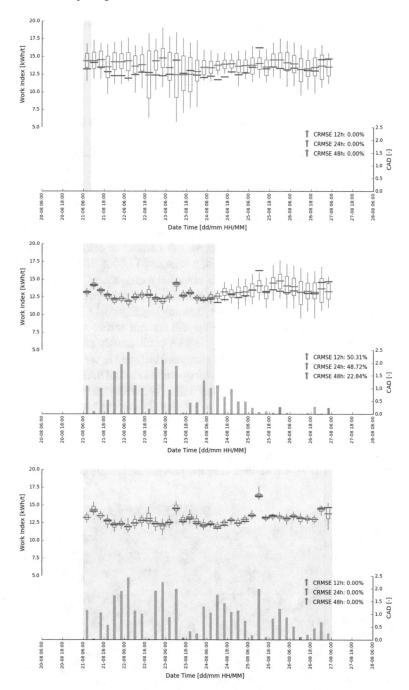

Fig. 4.7 Improvement of Bond Working Index prediction by continuously updating of the grade control model (top—not updated, middle—partly updated, bottom—fully updated)

4.2 Use Case 2: Updating Short-Term Quality Model Based on a Cross-Belt Scanner

4.2.1 Case Description

A resource-efficient extraction of mass mining products from an open-pit operation is a complex task with several, partly competing, objectives. For example, extracted material is to be delivered within a strictly and often contractual defined range of several quality-determining attributes. On the other hand, deposits are geologically complex in geometry and spatial distribution of elements and may consist of several seams or zones and splits, each having highly spatially varying quality parameters. A typical example of such types of deposits is iron ore or coal deposits. For the latter, product requirements are typically determined by strict upper or lower limits of various attributes, such as sulfur content, ash content or calorific value. These targets usually must be met on the basis of a train unit (1,000t–10,000t).

An essential basis for planning and operational management is a spatial model of coal qualities. In addition to planning decisions and the operational execution, modern operations conduct a dense quality monitoring. Figure 4.8 shows an example of a radiometric sensor for determining the ash content across a conveyor belt. A second example is the GNSS technology, which continuously records the position of excavators in an opencast mine. Applications can be found in various lignite mines, for example, the "satellite-assisted excavator operation control SABAS" system from RWE Power AG (Rosenberg 2007) or the online recording and analysis of actual cutting planes of the digging equipment at MIBRAG (Knipfer 2012).

The second use case demonstrates the opportunity to unlock the potential from operational monitoring in coal mines. It is based on a study conducted within the frame of the Real-Time Reconciliation and Optimization Project—RTRL-Coal (Benndorf et al. 2019). Cross-belt sensors provide information on coal quality

Fig. 4.8 Radiometric ash content sensor (*Source* MIBRAG mbH)

parameters in high spatial and temporal density, e.g., on the ash content of the coal extracted. To utilize this data for updating, following system elements are integrated:

- the short-term coal quality model, which is typically divided into digging slices,
- the observation model (material tracking model that combines GNSS data of digging and transportation equipment, time of extraction and conveyer belt speeds), and
- sensor data of the cross-belt scanner.

4.2.2   Input Variables

Next, the input is described in the context of coal quality model updating.

(a)   *Coal quality model*

Figure 4.9 shows the prior coal quality model of a particular bench, which illustrates the location of the coal seam within the bench slope and the spatial distribution of the ash content within the seam. Typically, these models are constructed from available exploration data using spatial interpolation techniques, such as linear interpolation. The application of the introduced Ensemble Kalman-Filter solution requires a set of geostatistically simulated resource models. To generate reliable simulation models requires some expert knowledge, which may not be available in an operational context. To overcome this hurdle, a more robust and simplified solution is investigated. Two approaches are compared:

(Approach 1) Base case—full simulation of geometry and coal quality using the Sequential Gaussian Simulation

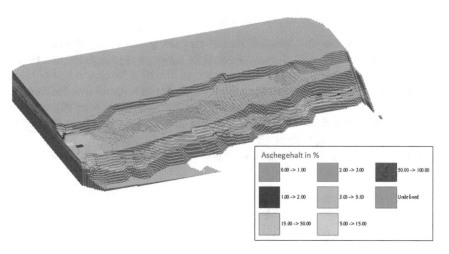

Fig. 4.9 Example of a cola quality model

(Approach 2) The proposed simplification generates the required prior model realizations by adding fluctuations around the company's short-term model. Typically, the latter is created by mining engineers or resource geologists, applying operationally suitable block geometries to the company's estimated grade control model. In this way, each block has an estimated coal quality value. In order to create the required realizations of the prior model, the following strategy is employed:

1. Short-term block model values are generally available in operations for each block and represent the prior estimation of block attributes.
2. A conditional simulation is applied to a production block support. A variogram model is used, which is derived from exploration data and regularized to the block geometry. Drill hole data with "zero values" are used as conditioning data while running the simulations around a zero mean. This step only requires the variogram and can be implemented as a "black-box" approach.
3. Simulated values of the production blocks from step (2) are added to prior estimations of block attributes from (1). The short-term model based on the simulations is now ready for updating.
4. The updated resource model (posterior model) is split in a mean part, which is written back to update the short-term block model (1). The uncertainty related part is written back in the ensemble part (2).

(b) *Observation or material tracking model*

With standard GNSS positioning systems used today in the mining industry, the location of extraction points, e.g., the bucket wheel, can be determined at each point in time with the necessary precision. Utilizing belt speed information, the position of the extraction equipment with respect to the belt, and the position of the cross-belt sensor, material-tracking information is available at high precision. This allows the direct assignment of sensor reading to the coal quality model.

(c) *Data from the operational monitoring*

For coal quality monitoring, standard sensors are used, which are installed above the conveyor belt (see Fig. 4.8). These can either be assigned to each extraction bench (one-to-one relationship) or installed above a central coal collection belt. The latter measures blended material from two or more benches (one-to-multiple relationship). For both cases, updating of the coal quality model is applicable (see Yüksel et al. 2017). In addition to these online collected data, also analysis data from train sampling can be used. For the unambiguous assignment of the sensor observations to the coal quality model, online data have to be provided with a synchronized timestamp.

4.2.3 Continuous Model Update and Results

Results from a one-to-one relationship (one sensor per bench)
Using model updating, online measurements are integrated into the coal quality model in near real time. Sensor data of coal qualities of an extracted block are filtered (cleaned for outliers) and fed back into the local coal quality model. Due to the spatial correlation to the neighboring blocks, the production forecast, especially for the next digging blocks, improved significantly. Figure 4.10 illustrates some results of this case study. The figure shows a comparison between prior model (prior model average), actual sensor data (black squares), and updated posterior model (light gray lines, the solid line shows the expected value). There are significant improvements in the prediction for the next extraction blocks, which are in the order of 65%. An evaluation of results suggests that the update is effective for about seven subsequent blocks. This roughly corresponds to the radius of influence of the information as determined by the variogram. This range applies to both directions, direction of digging, and direction of mining, and thus can significantly improve the medium-term forecast for the next few weeks and months. A detailed description of the case study is given in (Yüksel and Benndorf 2016).

Results from a one-to-two relationship (one sensor characterizes blended material from two benches)
In this case, the sensor is installed above a central conveyor belt and characterizes a blend of material originating from two extraction points. Figure 4.11 shows the application case using a fully geostatistically simulated prior model (Approach 1) and Fig. 4.12 for the short-term model (Approach 2). Both figures show the prior prediction of the ash content, sensor data from an RGI sensor, the expected posterior mean of the ash content, and associated error bars.

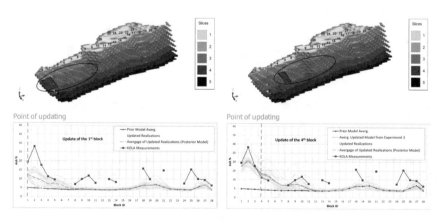

Fig. 4.10 Updating the prediction of the ash content of an open-pit rotary excavator using online quality data

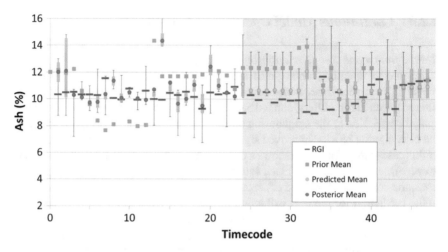

Fig. 4.11 Results based on conditional simulation: The green area represents the prediction period. The white area represents the learning period

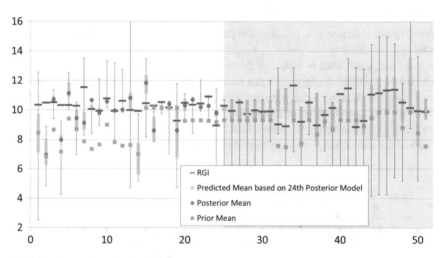

Fig. 4.12 Results based on the short-term model: The green area represents the prediction period. The white area represents the learning period

For both approaches, the relative improvement in the prediction of ash content is in the order of 10–35%. The range of improvements lasts for about 2–3 working days of excavator progress. This is due to the spatial correlation of the coal quality parameter. Blocks, which are mined in 4–5 days, will not be affected by updating. Here also, it is expected that there is a further improvement in medium term, as blocks are also updated in the mining direction, not only in the digging direction of the excavator.

Comparing both prior models, it is first to mention that they show on this local scale significant differences. This is mainly due to the level of information used to build these models. The simulated model is based on drill hole information only and simulated on a dense grid. The prior model from the short-term plan is based on block estimates, which have been generated including additional to drill hole data also operational data, e.g., from face mapping or face samples. In front of this background, it is quite remarkable that updating for both models performs extremely well. This demonstrates the self-learning ability of the ability and the fact that the choice and quality of the prior model are of lesser significance. The model picks up the true structure after a few iterations of updating (Yüksel et al. 2017). This effect shall be further evaluated in the third case study.

4.3 Use Case 3: Updating Mineral Contents of Blast Blocks in a Polymetallic Underground Mine

4.3.1 Case Description

This case study extends the cases presented so far to an underground setting. The short-term planning and production control process considered here is discontinuous and subdivided into four steps (Fig. 4.13):

- exploration and creation of a grade control model,
- block classification (classification of a blast block in types "waste material", "ore type 1", "ore type 2", etc.) by face mapping,
- short-term planning, and
- material logistics and operations control.

In each of these process steps, decisions have to be made with respect to control variables of the subsequent process step. Also, operative monitoring is implemented, mainly by grade control sampling and face mapping of the next blast block. This data,

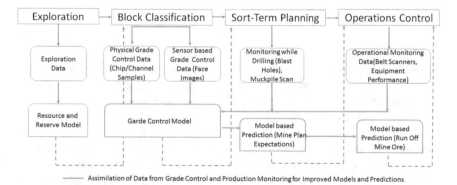

Fig. 4.13 Process and information flow for an underground application of grade control model updating in underground mining

together with information from the global resource model, is integrated into a short-term model that represents and summarizes the current most up-to-date information and thus serves as decision support. These models include the grade control model with the distribution of the ore properties within the production block, the muck pile model with the distribution of the ore and the grain sizes in the blasted heap and the logistic model with the properties of the ore along the entire logistic chain. The goal in this use case is to keep these models up-to-date by integrating all relevant operational information. For updating, the following three system elements are integrated:

- the short-term planning model (grade control model) with block geometry, geometry of the vein, and spatial distribution of the minerals within the vein,
- the material tracking information of the extraction and transport system and
- sensor data from face mapping, muck pile sampling, and further operational monitoring.

The use case, the Reiche Zeche mine, is located in Freiberg, Germany. Currently, it serves for research purposes and has been mined for Ag and for Cu, Pb, As during the past centuries. The Freiberg ore vein network is characterized by two shear systems, and spatially associated fissure veins. The Freiberg polymetallic sulfide deposit was formed by two hydrothermal mineralization events of Late-Variscan and Post-Variscan age. The ore minerals typical for this area are arsenopyrite, pyrite, sphalerite, chalcopyrite and agglomerated of galena-quartz-carbonate gangue (Stockmann et al. 2013). The test site is the "Wilhelm Nord" part of the "Wilhelm Stehender" orebody, located in the third level of the mine, at 200 m depth. The orebody itself is a continuous vein that has a dip of 50° ENE-WSW and a thickness that varies from 0.0 to 1.4 m.

To implement and benchmark the data assimilation process within a fully known environment, a synthetic data set based on real information is created. It uses data from a sampling campaign carried out between 1985 and 1989.

4.3.2   Input Variables

(a)   Grade control model

The grade control model is built based on data from a face sampling campaign on a 60 m × 60 m sized short-wall panel. Grade control data represent channel samples taken around the panel prior to extraction. The excavation panel is further subdivided into 2 m high extraction blocks of horizontal dimension 2 m × 2 m. Based on common geostatistical methods, the geometry of the vein and the spatial distribution of the minerals in the vein are simulated. Figure 4.14 shows a cross section and a top view of the vein. For each, the amount of ore, waste, and its mineral composition are predicted. To analyze the performance of the sequential updating algorithm applied to different grade control strategies, different scenarios constrained to different levels of information are evaluated.

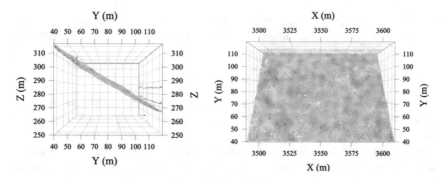

Fig. 4.14 Cross section (x–z-plane) and top view (x–y-plane) of the vein with the spatial distribution of the sphalerite content

- *Sampling strategy 1*: the first scenario mimics a sampling campaign, where grade control data are available around the full contour of the mining panel considered. The prior model generated based on this sampling campaign is denoted as Mine-Face scenario. Sampling information considered is represented in the right part of Fig. 4.15.
- *Sampling strategy 2*: the second scenario combines information from the Mine-Face scenario with data from three boreholes that are oriented towards the mining direction, which explore the center of the mining panel (left-hand side of Fig. 4.15) This scenario is referred to as Bore-Hole scenario.

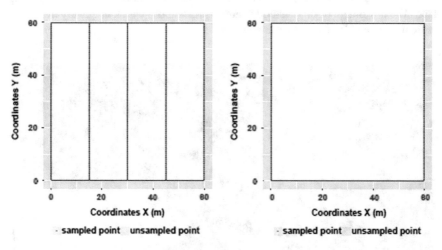

Fig. 4.15 Representation of the information accounted for in the sampling campaigns. The right part represents the mine-face sample information (sampling strategy 1) and the left part of the bore-hole sample information (sapling strategy 2)

- *Sampling strategy 3*: No geo-referenced grade control data are considered. The prior model is generated as unconditional simulation based on univariate statistics inferred from grade control data.

(b)   Observation/material tracking model

Depending on the position of the data acquisition in the extraction process, different observation models are used. For hyperspectral images, georeferencing in the local coordinate system is necessary to assign pixel information directly to the related part of the extraction block. Data collected from the muck pile or along the mining process can be related to the mining system by material tracking systems using underground positioning and tracking systems, such as B. RFID tags (Wortley et al. 2011).

(c)   Data from the operational monitoring

Face mapping by hyperspectral technology: Using a hyperspectral camera and suitable classification algorithms, ore at a working face can be classified in ore types or minerals. Instead of three channels (red, green, blue), a large part of the spectral signature of the ore is recorded pixel by pixel. This signature depends on the material itself as well as the lighting conditions underground. As a result, a spectrogram is available for each pixel of the hyperspectral image, which is characteristic of the mineral (Fig. 4.16, right corner). Classification algorithms help to assign a mineral class to each pixel. A proof of concept for practical underground use has been the subject of research in the projects Real-Time Mining (Desta and Buxton 2019) and ARiDUA (Varga 2016). Counting pixels for a particular mineral class allows to estimate the proportion of this mineral within the exposed ore vein at the working face. The advantage of this method is areal information, which is quickly available, without the need for laboratory analysis.

Fig. 4.16 Classification of the vein by hyperspectral technique and support vector machines, spectral signature of pyrite (after Varga 2016)

Point sensors in the extraction process: Point sensors can provide information on the composition of the ore at individual sampling points. Mobile X-Ray Fluorescence (XRF) spectrometers or LIBS (Laser-Induced Breakdown Spectroscopy) sensors that analyze the chemical composition are in the project. The latter has been developed as a prototype within the project Real-Time Mining by Spectral Industries (Dalm and Sandtke 2019). Applying this technique for face sampling, muck pile sampling, or conveyor chain scanning, provides temporally and spatially dense information on the chemical and mineralogical composition of the ore.

4.3.3 Continuous Model Updating and Results

The updating algorithm is applied to a part of the mining panel containing a total of 225 SMU's that represent 15 short-wall drifts with 15 SMU's per drift. The mining directions are upward similar to shrinkage stoping. The updating sequence per drift is defined from left to right. This direction is applied to each drift of the upcoming drifts (Fig. 4.17).

To update locally, a localization function is implemented that exponentially decreases the correlations with the distance to the block being currently extracted (e.g., Leeuwenburgh et al. 2005). This function has a radius of 8 m from the center of the SMU's that is being updated. Further than this range, blocks are not considered for updating. The application of this gradually decreasing function guarantees that there is no abrupt change of domain between the updated and non-updated areas.

Figure 4.18 shows the box plots of simulate realizations of As values per SMU in the ninth drift unconditioned to sampling data (sampling strategy 3). The red point plotted in each SMU corresponds to the true SMU As grade. The upper figure shows the model before the assimilation. The lower figure shows the prediction after assimilating until SMU 140. There is a marked reduction in the uncertainty of the after assimilation. The SMU means for SMUs 136–140 are now very close to the respective true means. The predictions of SMU141 ff demonstrate a significant decreased prediction error. Figure 4.19 shows a similar illustration for sampling strategy 2.

Figure 4.20 shows the mean average error of the prediction of the next SMU to be mined as an average value over the whole panel before and after the assimilation process from right to left: the unconditional scenario (sampling strategy 3) the mine-face scenario (sampling strategy 1) and the bore-holes scenario (sampling strategy 2). Looking at non-updated models, as expected, the error for the unconditional model is higher than for the conditional to bore-hole and mine-face information. Also, one can see that the error decreases with the amount of conditioning information available, therefore the bore-hole scenario has a smaller error than the mine-face scenario. On the other hand, considering the updated models, only minor differences in their mean absolute errors can be observed between the different scenarios. The mean absolute error decreases significantly for all three scenarios. Interestingly, conditioning the different grade control models to the production information overwrites the initial conditioning information quickly. This indicated that the self-learning ability of the

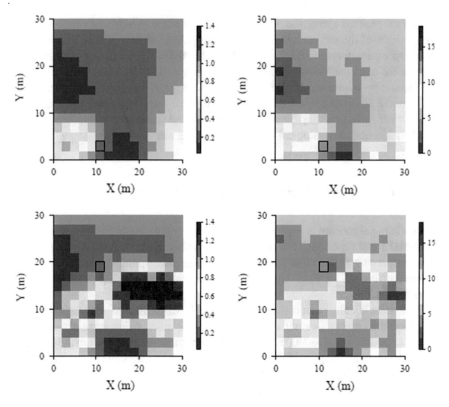

Fig. 4.17 Grade control model of the vein thickness (left) and the sphalerite content (right) at different points in time of extraction

updating framework kicks in quickly after starting the sequential assimilation process. The initial level of information with respect to grade control information of the mining panel is of less significance. A more detailed discussion on this case can be found in Prior et al. (2020a).

4.3.4   Practical Aspects

The previous case studies demonstrate both, the applicability on an industrial scale and also the added value. Next, few experiences and practical implementation considerations shall be discussed.

Availability of a fully reconciled grade control model: The updating framework improves the precision of forecasting block values related to attributes of interest, such as the BWI or coal qualities. In addition, blocks that have been mined in the past are updated leading to a fully reconciled block model. The availability of such a model allows to conduct more detailed studies, especially on the relation between

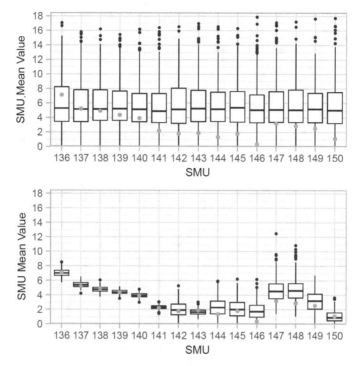

Fig. 4.18 Prior prediction of the As values for drift 9 (upper part) and updated prediction after assimilating values until block 140 (lower part) for the unconditional scenario (sampling strategy 3)

resource model, prior grade control model, and reconciled grade controlled model. This enables improving resource models based on short-term learnings.

Observations on blends of material: As demonstrated in case studies one and two, also observations on blends on materials lead to local improvements of the grade control models. Due to the high frequency of online sampling, the algorithm can solve existing ambiguities. Depending on the sampling frequency, also cases with more than two sources of materials for a scanned blend can be applied to updating. For more information, the reader is referred to Wambeke et al. (2018).

Quality of the prior model of less significance: As demonstrated in case study two and three, the quality of the prior model is of less significance on the performance of the updating algorithms. This implies that even if less information are available about a mining panel, as soon as the assimilation process is started, a self-learning ability kicks in quickly and the learnings from process monitoring overwrite prior information. If this observation can be substantiated by other use cases and over a longer time span, an optimization (meaning a reduction) of prior sampling/exploration can be considered.

Flexibility in the observation model: As demonstrated in all three case studies, production data used to update the grade control model are only indirectly linked to

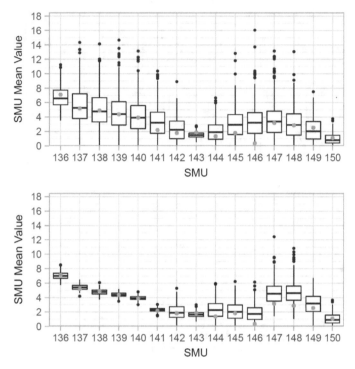

Fig. 4.19 Prior prediction of the As grade for drift 9 (upper part) and updated prediction after assimilating values until block 140 (lower part) for sampling strategy 2

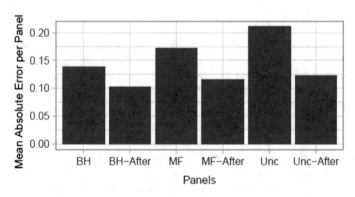

Fig. 4.20 Mean average error for each of the three different sampling strategies before and after updating evaluated at all the SMU's that are one block-distance away from the SMU being currently extracted

the grade control model itself. In fact, there is no analytical expression that describes this relation. Observations from production monitoring and the grade control model are linked by a general observation model, which can be a material tracking system or a process simulation. The ensemble approach in updating allows for this flexibility and makes the updating approach very flexible in implementation. It has the ability to utilize models and software tools, which are often already in place but currently not linked together, such as material tracking or monitoring systems. The integrative character of the updating framework combining grade control models, material tracking systems and production monitoring, is attractive for data fusion solutions and provides a concrete solution to the digitization initiatives in the mining industry.

Model updating is based on possibly biased sensors: Care has to be taken when overwriting prior modes based on analytical sample data with observations from online monitoring sensors. These often deliver indirect data of lower precision and have to be regularly calibrated for use. They may underlie a timely dependent drift component that introduces a systematic error or bias. For example, in use case 1, the ball mill may deliver other sensor responses after a major overhaul. For a routine application of updating algorithm, sensors should be calibrated at regular intervals.

Localization function: The ensemble approach combined with the empirical determination of covariances with a limited number of ensemble members may lead to nonrealistic updates in remote parts of grade control model. A localization function, as described in use case three, will limit the updating on the local neighborhood, from which the raw material being characterized by monitoring is originating from.

Timely resolution of the transportation model: The updating intervals mainly depends on the ability to link the material being observed to its origin in the deposit. This ability is mainly determined and limited by the operational material tracking. Even, when using online tracking systems, such as GNSS positioning of trucks, the utilization of ROM fingers or stockpiles weakens the ability of precise tracking. This limits updating intervals to longer timespans. In many operations, daily updating is realistic. If updating is intended to be intensified, then a more detailed material tracking may the solution.

Unique time stamp and coordinate system: Updating links the grade control model, observations from sensor stations and the material tracking system. This requires discipline on location and time tracking and synchronized systems in time and geodetic coordinate system.

Geo-database: The main experience in building up the case studies has been, that establishing a consistent database, which can be used for updating, required a large timely effort. Once established, the framework can be rather easily applied. If mining operations are well set up in geo-data management, necessary databases can be easily accessed. A good consistent geo-database and data management are the main enablers for updating, but also for any effort in digitizing processes.

Further applications: While writing this book, further areas have been developed in resource and grade control model updating. To name just a few, updating based on compositional data (Prior et al. 2020b) and updating destination policies based on re-enforcement learning (Paduraru and Dimitrakopoulos 2019). While this book aims to provide an introduction to the topic, it is expected that more sophisticated methods and relevant applications will be published in the near future.

References

G.R. Ballantyne, M.S. Powell, Benchmarking comminution energy consumption for the processing of copper and gold ores. Miner. Eng. **65**, 109–114 (2014)

J. Benndorf, R. Donner, M. Lindig, O.M. Lohsträter, H. Rosenberg, S. Asmus, W. Naworyta, M.W.N. Buxton, Real-time reconciliation and optimization in large open pit coal mines (RTRO-Coal)–Final report. European Commission, Directorate-General for Research and Innovation, Directorate D-Industrial Technologies, 112 p. (2019). https://doi.org/10.2777/61704

M. Dalm, M. Sandtkem, Geochemical mapping of drill core samples using a combined LIBS and XRF core scanning system, in *Real Time Mining-2nd International Raw Materials Extraction Innovation Conference in Freiberg, Germany*, ed. by J. Benndorf (2019). ISBN 9783938390429

F.S. Desta, M.W.N. Buxton, Chemometric analysis of mid-wave infrared spectral reflectance data for sulphide ore discrimination. Math Geosci **51**, 877 (2019). https://doi.org/10.1007/s11004-018-9776-4

R. Donner, M. Rabel, I. Scholl, A. Ferrein, M. Donner, A. Geier, A. John, C. Köhler, S. Varga, Die Extraktion bergbaulich relevanter Merkmale aus 3D-Punktwolken eines untertagetauglichen mobilen Multisensorsystems, in *Proceedings Geomonitoring 2019* (2019), pp. 91–110. https://doi.org/10.15488/4515

A. Knipfer, Decision support for operating lignite mines with high geological variability based on mine surveying and production monitoring. Bergbau Zeitschrift für Rohstoffgewinnung Energie Umwelt **4**, 148–152 (2012). (In German)

O. Leeuwenburgh, G. Evensen, L. Bertino, The impact of ensemble filter definition on the assimilation of temperature profiles in the tropical Pacific. Q. J. R. Meteorol. Soc. J. Atmos. Sci. Appl. Meteorol. Phys. Ocean. **131**(613), 3291–3300 (2005)

A. Lynch, A. Mainza, S. Morell, Ore comminution measurement techniques, in *Comminution Handbook*, ed. by A. Lynch. Spectrum Series, vol. 21 (The Australian Institute of Mining and Metallurgy, Carlton Victoria, Australia, 2015), pp. 43–60

C. Paduraru, R. Dimitrakopoulos, Responding to new information in a mining complex: fast mechanisms using machine learning. Min. Technol., 1–14 (2019)

A. Prior, J. Benndorf, U. Müller, Resource and grade control model updating for underground mining production settings. Math. Geosci. (2020a) (in print)

A. Prior, R. Tolosana-Delgado, K.G. van den Boogaart, J. Benndorf, Resource model updating for compositional geometallurgical variables. Math. Geosci. (2020b) (accepted)

H. Rosenberg, Detection of materials and deposits as a basis for innovative operations management systems employed as part of opencast mine process optimizations. World Min. Clausthal Zellerfeld **59**, 173–180 (2007)

M. Stockmann, D. Hirsch, J. Lippmann-Pipke, H. Kupsch, Geochemical study of different-aged mining dump materials in the Freiberg mining district, Germany. Environ. Earth Sci. **68**(4), 1153–1168 (2013)

S. Varga, Multisensorsystem für die automatisierte Detektion von Gangerzlagerstätten und seltenen Erden in einer Mine. Schriftenreihe des Institutes für Markscheidewesen und Geodäsie der TU Bergakademie Freiberg, Heft 2016-1 (Wagner Digitaldruck und Medien GmbH, Nossen, 2016), pp. 276–283. ISBN 978-3-938390-17-7

T. Wambeke, D. Elder, A. Miller, J. Benndorf, R. Peattie, Real-time reconciliation of a geometallurgical model based on ball mill performance measurements–a pilot study at the Tropicana gold mine. Min. Technol. (2018). https://doi.org/10.1080/25726668.2018.1436957

M. Wortley, E. Nozawa, K.J. Riihioja, Metso SmartTag–the next generation and beyond, in *35th APCOM Symposium* (2011, September), pp. 24–30

C. Yüksel, J. Benndorf, Performance analysis of continuous resource model updating in lignite production, In *Valencia Geostatistics 2016*, eds. by J. Gómez-Hernández, J. Rodrigo-Ilarri, M.E. Rodrigo-Clavero, E Cassiraga, J.A. Vargas-Guzmán, vol. 2 (Springer, 2016), page 65 ff

C. Yüksel, J. Benndorf, M. Lindig, O. Lohsträter, Updating the coal quality parameters in multiple production benches based on combined material measurement: a full case study. Int. J. Coal Sci. Technol. **4**(2), 159–171 (2017)

Chapter 5
Optimization Methods to Translate Online Sensor Data into Mining Intelligence

Abstract Online sensor data from production monitoring deliver a continuous database and up-to-date information about the characteristics of the produced ore. The updating framework presented in Chaps. 3 and 4 enables to manifest this data into up-to-date knowledge about the orebody. The final step in the closed-loop approach is to translate this up-to-date knowledge into intelligent decisions for short-term planning and production control. This Chapter first briefly introduces general aspects of mine planning optimization. Two examples describe case studies of implemented short-term mine planning optimization that take updated grade control models into account. An attempt to quantify the added value of information from production monitoring conclude this Chapter.

5.1 Introduction to Mine Planning Optimization

Mathematical Optimization or Operation Research formulates the task to be solved as an extreme value problem with boundary conditions. It is composed of the following constituents.

(1) Decision variables **x**. A typical decision variable in mining is a variable x^t, that assigns an extraction period t to a smallest mineable unit x. The range of possible values of x^t is 0 or 1. Is it 1, block x is extracted in period t, else not. Considering only a small grade control model consisting of 10.000 blocks to be scheduled over 5 periods requires already an order of about 50,000 decision variables. More complex cases add a destination policy, meaning that next to the time period the block destination (waste dump, costumer 1, costumer 2, etc.). is to be decided on. Of course, decision variables vary from application case to application case and have to be chosen accordingly.

(2) The objective function that establishes the relation between objective value z and decision variables $z = f(x)$, which has to be minimized or maximized. It defines the dependency of the objective value, e.g., operating costs or NPV, from decision variables. To establish this relation, parameters are necessary, which characterize properties of mining units. A typical example is the expected

monetary value of an SMU, which is mined in time period t. These parameters can be derived from geostatistically estimated grades of the block.

(3)   A system of equations that describe the boundary conditions or constraints as a function of the decision variables $g(x) <=> b$. Boundary conditions limit the feasible solution space. Typical constraints include upper and or lower limits of production capacities per time period, minimum and maximum grade of elements or precedence relationships in the production sequence due to geotechnical considerations (slope constraint).

The optimization problem is summarized as the following set of equations:

$$\text{Maximize/Minimize} \quad z = f(x_1, x_2, \ldots x_n)$$

$$\text{Subject to} \quad g(x_1, x_2, \ldots x_n) \quad \begin{Bmatrix} \leq \\ \geq \\ = \end{Bmatrix} \quad b_i \quad i = 1, \ldots m$$

$$x_j \geq 0, \quad j = 1, \ldots, n$$

Large models can be characterized by several millions of decision variables and constraint equations. Depending on the size, complexity of the objective function (linear or nonlinear) and the type of decision variables (discrete or continuous) and parameters of the boundary conditions (deterministic or stochastic), many different solution algorithms can be used. These include linear programming, dynamic programming, integer programming, stochastic programming, metaheuristic techniques such as ant colony optimization, simulated annealing, or genetic algorithms, just to name a few. A more complete overview, description of the individual approaches and their limits can be found in the standard literature (e.g., Winston and Goldberg 2004).

5.2   Optimization for Short-Term Planning and Production Control

To investigate the expected benefit of the updating framework, it is applied in the short-term mine planning context. To cover different aspects of the short-term mine planning horizon, two different simulation-based optimizers are considered, which have been implemented within the context of the European Project RTRO-Coal. Both models complete a particular solution for the closed-loop approach (Table 5.1).

The weekly job scheduling optimizer considers shift-based scheduling (a shift being 8 h long) and it shows the benefits of the updating framework—to quickly integrate newly gained information into production control. In contrast, the short-term optimizer for extraction sequencing is used to demonstrate the benefits on a broader, less localized, scale and a longer time frame of weeks to months.

Table 5.1 Characteristics of the short-term and weekly job scheduling mine planning optimization models

Model	Short-term mine planning	Weekly job scheduling
Task	Monthly mine plan	Weekly/daily job schedule of excavators
Approach	Mixed-integer programming to model the mining operation and determining the best configuration for a set of decision variables	Discrete event simulation to simulate the mining operation in combination with a hybrid genetic and simulated annealing search algorithm determining the best configuration for a set of decision variables
Objectives	Meeting weekly coal tonnages and grade targets	Meeting daily coal tonnages and grade targets
Constraints	• Planned equipment's maintenance schedule (equipment availability) • Equipment capacities (production rates) • Precedence relationships within a bench • Required equipment downtimes while belt shifting occurs	• Planned equipment's maintenance schedule (equipment availability) • Equipment capacities (production rates) • Resources: number of available crews • Extraction sequence in different benches
Decisions	• Extraction sequence in different benches • Belt shift scheduling • Blending of coal product	• Excavator task scheduling • Blending of coal product
Time steps	Weeks	Production shifts

5.2.1 Short-Term Mine Planning

The short-term optimization model investigated here is based on mixed-integer programming and considers a continuous lignite mining operation.

$$min\{c^T x | Ax \leq b, x \geq 0, x_i \in \{Z, R\}\} \tag{5.1}$$

The mining process is represented by a system of linear (in-) equalities $Ax \leq b$ and a set/vector of decision variables x defining a mining plan. These linear inequalities are also referred to as constraints, and consider limits imposed by the operation such as the amount of extractable material per time period due to the deployed equipment. Each decision variable x_i can either be binary (to determine the extraction sequence, Table 5.2) or continuous (to determine coal blending and when a belt shift should occur).

Which mine plan is the best is determined by the objective function $c^T x$. For short-term mine planning, deviations from production targets shall be minimized when excavating material from the mine according to the plan. Production targets are in terms of coal tonnage and coal quality (e.g., ash content) to ensure the reliable

Table 5.2 Illustration of how the binary decision variables determines the extraction sequence. While block 1 will be mined in extraction period 3, block 2 is not scheduled for mining during extraction periods 1 through 4

Block	Extraction period			
	1	2	3	4
1	0	0	1	0
2	0	0	0	0

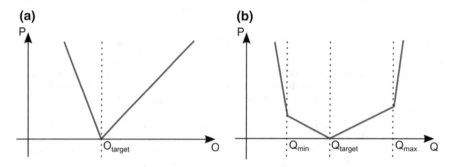

Fig. 5.1 Penalty functions for not meeting production targets: **a** coal tonnage and **b** coal quality (reproduced after Yüksel et al. 2019)

and continuous delivery of in-spec coal to the customers. Therefore, penalty functions for not meeting production targets have been defined (Fig. 5.1).

Figure 5.1a shows a steeper slope for underproduction in comparison to overproduction since in this instance there is not enough lignite provided to the customer, e.g. endangering the correct operation of a power plant. In contrast, overproduction is not viewed as critical as underproduction because the considered mining operation has a lignite bunker at its disposal to temporarily store a limited amount of lignite to compensate for such things or standstills due to holidays, equipment failure, etc.

The penalties for deviations in coal tonnage from the production target O_{target} can be calculated in a similar way as the coal quality penalties, which are explained in detail next.

For the coal quality production target (ash%), the penalty function is a linear piecewise function consisting of four segments with slopes sl_1, sl_2, sl_3, and sl_4 (€/ash% · t). The lower and upper limits of a coal quality parameter Q_{min} and Q_{max} are taken into account, which represent the maximum bandwidth for an efficient operation of the downstream process, such as power plants. Exceeding these thresholds will be penalized stronger than just slightly deviating from the target value Q_{target}.

For the penalty calculation Eq. (5.2) holds, where Q is the simulated quality.

$$P = \begin{cases} sl_1 * (Q - Q_{min}) + sl_2 * (Q_{min} - Q_{\text{target}}) & Q \leq Q_{min} \\ sl_2 * (Q - Q_{\text{target}}) & Q_{min} < Q \leq Q_{\text{target}} \\ sl_3 * (Q - Q_{\text{target}}) & Q_{\text{target}} < Q \leq Q_{max} \\ sl_4 * (Q - Q_{max}) + sl_3 * (Q_{max} - Q_{\text{target}}) & Q_{max} < Q \end{cases} \quad (5.2)$$

To incorporate the resource model uncertainties, a neutral risk approach is selected by calculating the penalty value for each simulated quality value of the resource model and then determining the mean of these penalty values $\bar{P}$. For the posterior model, this calculation is done in a similar manner.

5.2.2 Weekly Job Scheduling

Due to the complexity of interacting constraints in production scheduling, for weekly job scheduling a flexible approach based on discrete event simulation and a hybrid genetic and simulated annealing search algorithm is implemented. A simulation model represents the complex continuous mining operation. This simulator serves as the objective function to evaluate different mining plans (value combinations of the decision variables x). To find the best mining plan, the simulator interacts with an optimizer. The optimizer explores the space of the decision variables and suggests a mining plan that is evaluated by the simulator. Based on the evaluation results, the optimizer chooses which mining plan to evaluate next (Fig. 5.2).

For the weekly job scheduling process, the inputs/constraints to the simulator include but are not limited to the resource model, a given extraction sequence (mining plan), and the equipment's production rates and scheduled maintenance requirements. The decision variables of the optimization process are the task schedules of the excavators. The schedule for each excavator is a list with three shifts per day of the simulated period. For each shift, the excavator is either scheduled to work or not active. The input for a single simulation is thus a two-dimensional array with six rows and $3 \times n$ columns, where n is the number of simulated days. An example of a schedule is shown in Fig. 5.3 (Mollema 2015).

The stochastic mine optimizer also works with a penalty function to calculate the fitness of a solution. The total penalty value of a solution is the sum of both, the quality

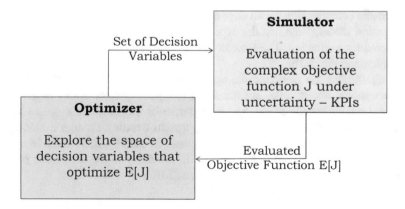

Fig. 5.2 Interaction between an optimizer and a simulator (reproduced after Benndorf et al. 2015)

Mo			Tue			We			Thu			Fri			Sat			Sun			
X	X	X			X	X	X	X	X	X	X	X	X	X	X	X	X				Excavator 1
X	X	X	X	X	X	X	X	X			X	X	X	X							Excavator 2
	X	X		X	X		X	X		X	X		X	X		X	X		X	X	Excavator 3
X	X	X	X	X	X	X	X	X	X	X	X	X	X	X	X	X	X	X	X	X	Excavator 4
	X			X			X			X			X			X			X		Excavator 5
												X	X	X	X	X	X	X	X	X	Excavator 6
X	X	X	X	X	X	X	X	X	X	X	X	X	X	X	X	X	X	X	X	X	Excavator 7

Fig. 5.3 Visual representation of a series of task schedules, with 7 simulation days and 3 shifts for each day. A cross means the excavator is scheduled to work (after Mollema 2015)

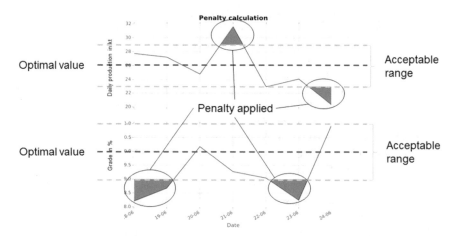

Fig. 5.4 Penalty calculation for a job schedule

and tonnage penalty value for all days of the simulation, similarly to the short-term optimization model. Note that the expected penalty over all ensemble members is considered for evaluation. The structure of both penalty functions is similar to the one depicted in Fig. 5.1b.

By minimizing the penalties, the optimizer tries to find the best task schedule. Resource model uncertainties are translated into penalty calculations based on an optimized mining schedule (Fig. 5.4). This particular implementation uses a hybrid simulated annealing—genetic algorithm approach. Iterating through a number of optimization cycles, the penalty value decreases. Eventually, this results in a close to optimal schedule (Fig. 5.5). For detailed information on the used mine process optimizer, the readers are referred to Mollema (2015).

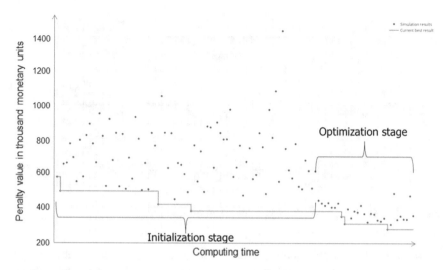

Fig. 5.5 Development of the penalty value during the simulation based optimization process

5.3 Value of Information

To investigate the impact of the improvements achieved by the closed-loop frame-work, next an attempt to quantify the economic impacts in mine planning is made.

One of the commonly used tools for assessing the value of additional information added into a system is the Value of Information (VOI) (Howard 1966). In the past decades, VOI gained high popularity in many different fields. A few applications also appeared in the mining industry. Peck and Gray (1999) make no explicit reference to VOI, yet they discuss the potential benefits for decision-makers by gathering information in the mining industry. Barnes (1986) applied VOI to incorporate geostatistical estimation into mine planning. Eidsvik and Ellefmo (2013) conducted a VOI study in order to compare two grade-analyzing methods (differing in costs) in the collected data from exploration boreholes. Eidsvik et al. (2015) presented a unified framework for assessing the value of potential data gathering schemes by integrating spatial modeling and decision analysis, with a focus on petroleum, mining, and environmental geosciences. Contrary to the mining industry, the VOI approach has found more applications in a related field, the oil and gas industry. Bratvold et al. (2009) provided an extensive overview of oil and gas industry applications. Barros et al. (2016) proposed a new methodology to perform VOI analysis within a closed-loop reservoir management framework, which is a similar framework of resource model updating.

The essence of the technique is to evaluate the benefits of collecting additional information before making a decision (Bratvold et al. 2009). When using the updating framework, the decision-making process would change the short-term mining plan by using a mine optimizer. If the resource model always provides correct coal

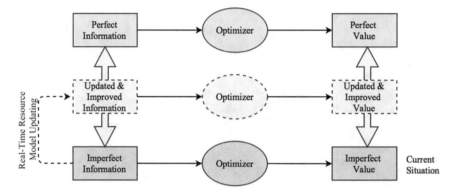

Fig. 5.6 Value of Information Concept within the closed-loop approach

quality attributes, it would deliver Perfect Information (PI), otherwise it is known as Imperfect Information (II). The latter is usually the case in geoscience applications, since the reality is unknown. The resource model updating framework aims to carry forward the current situation from Imperfect Information to an "Improved" Imperfect Information state, where the current situation lies somewhere between the Perfect Information and previous Imperfect Information (Fig. 5.6). The previous Imperfect Information state gets closer to the Perfect Information state with each iteration step of the updating process.

The expected benefit of additional information (integration of the online sensor measurements into the resource model) is compared to a case where there is no additional information. These benefits are evaluated based on the economic impact, for example, the monetary value such as cost/penalties per shift of mining operation, determined by applying the resource model updating framework in the mine planning process. In general, VOI is calculated in the following manner:

$$VOI = \begin{bmatrix} \text{Expected value with} \\ \text{additional information} \end{bmatrix} - \begin{bmatrix} \text{Expected value without} \\ \text{additional information} \end{bmatrix} \quad (5.3)$$

In the coal-based case example presented later, the concept analyzes the value of the resource model updating framework's ability to improve the prediction of the ash percentage (ash%). For this, the expected value of the posterior model ($V_{posterior}$) is compared to the prior model's expected value (V_{prior}).

$$VOI = V_{posterior} - V_{prior} \quad (5.4)$$

Translated into economic terms the VOI (Eq. 5.5) of the resource model updating framework can be expressed as

$$VOI_{economical} = |C_{posterior} - C_{prior}|. \quad (5.5)$$

The performed research focuses on the costs of deviating from the target quality (ash%) during short-term production. The calculation of these costs follows Eq (5.6). C_{prior} (€) is the costs of deviating from the target production quality, when executing the mine plan on the prior model. The unit costs for deviating per ton of coal is D_{prior} (€/ash% · t). The amount of the deviation per coal quality value is d_{prior} (ash%) and finally, the amount of the deviated coal is t_{prior} (ton). Similar applies to the posterior model. The previously defined parameters are: $C_{posterior}$, $D_{posterior}$, $d_{posterior}$, and $t_{posterior}$, respectively. Then, the costs of deviating from the target production are

$$C_{prior} = D_{prior} * d_{prior} * t_{prior} \qquad (5.6)$$

The VOI concept considers the Value of Perfect and Imperfect Information (VOPI). Perfect Information refers to perfectly reliable information; thus it contains no uncertainties. Perfect Information rarely exists, but it provides a best-case scenario for the VOI and it defines an upper limit on the value of additional information (Phillips et al. 2009). Since the study presented here presents a real case, the reality remains unknown. Thus, there can be no VOPI defined. As a benchmark, a resource model is used that integrates all available sensor information available to the end of the study period. As indicated in Fig. 5.6, the performed experiments will compare the calculated VOIs of the Imperfect Value and Updated and Improved Value. These values will be calculated after applying the mine optimizers, described in the next section, on the prior and posterior (updated) models.

5.4 Case Study

5.4.1 Case Description

The case study is performed on a lignite mining operation in Germany, where the geology of the field is complex, including multiple split seams with a strongly varying seam geometry and coal quality distribution (Fig. 5.7). In this case study, the challenge originates from the complicated geology that leads to geological uncertainty associated with the detailed knowledge of the coal deposit. This uncertainty causes deviations from expected process performance and affects the sustainable supply of lignite to the customers. The knowledge of the coal deposit is improved and the process performance is increased by applying the resource model updating framework described previously. Now, the aim is to quantify the added value of the mentioned improvements in the knowledge.

For the case study, the target area is defined as an already mined out area of 25 km² with about 3000 drill holes. Mining operations are executed by six excavators, each working on a different bench. The produced materials are transported by conveyor belts. All conveyor belts merge at a central conveyor belt leading to the coal stock and blending yard, which is further connected to a train load station. A radiometric

Fig. 5.7 Complicated geology in the lignite mine Profen (Germany)

sensor measurement system is installed above the central conveyor belt just before the coal stock and blending yard. This system allows an online determination of the ash content of the blended mass flow directly on the conveyor belt, without requiring any sampling or sample processing. This case study assumes the RGI values to be correctly calibrated and representative.

The prior coal quality model is generated by 25 realizations using geostatistical simulations (SGS) on a $25 \times 25 \times 1$ m grid (SGS performed on each grid point) and afterward relocked to block size of about 50×50 m.

5.4.2   Applying the Model Updating Framework

The experiments performed in this case study calculate the expected values for differently informed resource models (Fig. 5.8). The base case resource model, namely the prior model, is without any additional information. The other resource models (posterior models) are built with sensor additional information. They are created by applying the resource model updating algorithm up to different points in time. In total, there are five different posterior models representing different levels of information.

This updating process is performed every 2 h using the actual mining sequence and the available RGI data for this time span. All of the updating parameters are kept constant in order to compare the differences caused by feeding different prior models as input. The posterior model, which resulted from updating the prior model over the July 19–August 13 period is assumed as the most precise model since this is

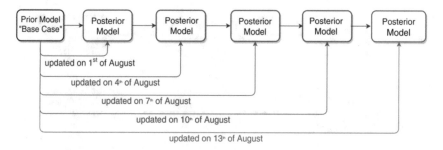

Fig. 5.8 Resource models that are used in the experiments

the most up-to-date resource model created with the highest amount of exploration data, and therefore serves as a benchmark.

Weekly Job Scheduling Mine Planning

The defined daily coal production is 12,000 tons and no penalties are applied for values between 9,000 tons and 15,000 tons, a deviation of 3,000 tons. The target production quality is defined as 9% ash and the penalty is only applied for the realizations above 10.5 ash% and below 7.5 ash%. The costs of deviating from the targets (the penalties) in this study are calculated by one unit per ton of coal. Hence, these penalties can be interpreted as a percentage of deviation from the targets. A penalty of 0.1€/ash% is applied per ton of coal. The mine planning optimizer is applied to all of the resource models, defined previously, for the following 5 days after their updating periods. For the base case and the benchmark case, the mine optimizer is applied for 5 days after each updating period. These dates are: August 2–6, August 5–9, August 8–12, and August 11–16. For these dates, there are different best schedules optimized based on the prior model, the updated model and the real model. Thus, in total there are twelve different best schedules.

Next, these obtained best schedules are applied to the benchmark model (Fig. 5.9). This is done in order to see the improvements during the mining operations, when using the prior model, the most current model and the benchmark model.

As mentioned already the case study presented here is a real case study and thus, there can be no VOPI defined for this case study. For this reason, the most precise model that incorporates most information is assumed to approximate reality. In this way, an approximation of VOPI can be calculated between the benchmark and the prior model whereas the VOI is calculated between the posterior model and the prior model.

A comparison of this would answer the following questions: "What would be the result of the mining activities if we didn't have additional information?", "How did the additional information affect the mining activities?", and finally "What would happen if we knew the reality and performed the mining activities mine based on that?" Of course, the latter one is only for comparison and in reality; we can never have this information beforehand.

Once the optimized schedules are applied to the benchmark model, the expected costs of deviating from the target quality (ash%) values are calculated. The expected

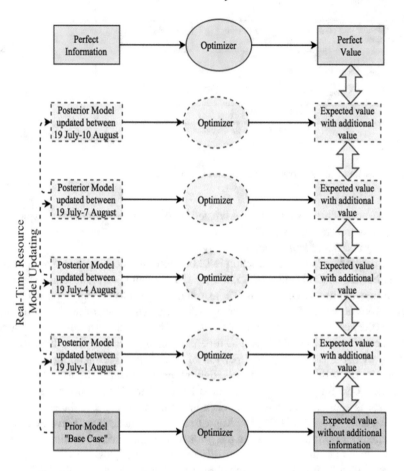

Fig. 5.9 VOI—Experimental scheme for weekly job scheduling mine planning

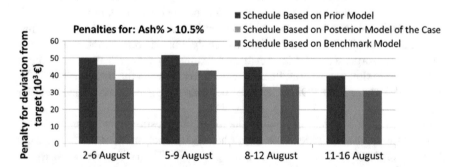

Fig. 5.10 Cost calculations of deviating from the target quality (ash%)

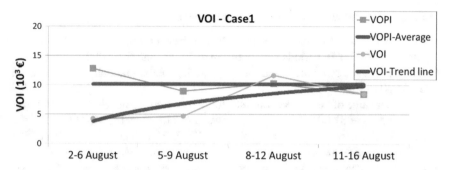

Fig. 5.11 Value of Information

values are then compared to each other and the VOI is calculated. This comparison and the results are provided next. Figure 5.10 presents the calculated deviation costs (penalties). This graph calculates the deviations per day for exceeding the upper target values of the ash content. Figure 5.11 presents the calculated VOI for each case. The VOPI is represented with squared lines and the average of those VOPI is represented with a red line. The calculated VOI is represented with a pointed line and the trend line fitted on these points is represented with a dark green line.

In Fig. 5.10, the dark gray column represents the calculated penalties for the optimized schedule based on the prior model. The light gray column represents the calculated penalties for the schedule optimized based on the posterior model, which is updated until the mentioned time period. The medium-dark gray column represents the calculated penalties for the schedule using the benchmark model. When evaluating the VOI for the next five days of mining after updating, a significant penalty reduction can be observed for the posterior model. This leads an increasing VOI towards VOPI.

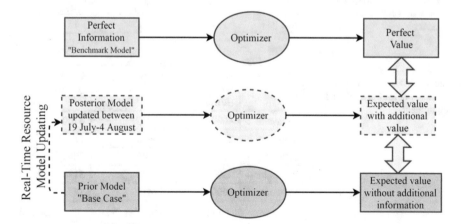

Fig. 5.12 VOI—Experimental scheme for short-term mine planning

Further, the following observations can be made:

- The mean value of penalties for the schedule based on the prior model over all periods is € 47,000, periods vary between € 40,000 and € 52,000.
- Penalties for the schedule based on the posterior model gradually decrease from € 46,000 to € 31,000.
- The mean value of penalties for the schedule based on the benchmark model over all periods is € 37,000, periods vary between € 31,000 and € 43,000.
- Thus, for this case study, the calculated VOPI is € 10,000 and the calculated VOI moves from € 4,000 to € 8,000 (Fig. 5.11) relating to a five day period.
- The above-mentioned VOI numbers will lead to approximately € 300,000– € 600,000 annual cost reduction or saving. Note that these savings are solely related to the weekly task scheduling application applied to the upper limit of ash content.

In Fig. 5.11, the trend line of the VOI illustrates the benefit of using a combination of the resource model updating algorithm and the mine optimizer ("closed-loop" optimization). With each iteration of updating, the mine schedule optimization penalties decrease, thus VOI increases.

Short-Term Mine Planning

Tables 5.3 and 5.4 summarize the optimization model's parameters used in this case study. The model starts with a predefined state of the open-cut mine and plans the exploitation on five benches for four time periods (weeks). 200 000 tons of lignite per time period (week) need to be supplied with a defined quality bandwidth for the ash content (Table 5.3).

To penalize exceeding or a shortfall with respect to defined lignite quality limits Q_{min} and Q_{max}, a slope coefficient (sl_1, sl_4) of "±1.3" is being used (Fig. 5.1b).

Table 5.3 Defined quality bandwidths for the lignite product

	Q_{min}	Q_{target}	Q_{max}
Ash [%]	0	10.7	15

Table 5.4 Information about the optimization model parameters for the deployed equipment

Excavator	Operating hours [h/week][a]	Waste [m³/h][a]	Lignite [m³/h][a]
Excavator 1	4000	7050	4700
Excavator 2	2800	3880	3170
Excavator 3	2300	1645	2000
Excavator 4	2500	2820	3170
Excavator 5	4000	1410	1530

[a]Please note that out of confidentiality reasons the presented parameters are scaled and do not represent reality

This represents 1.3 units of monetary value per unit deviation (€/ash% · t). Within the quality limits, a slope coefficient (sl_2, sl_3) of "± 1" is used.

For penalizing over- or underproduction, a slope coefficient of "40", respectively, "80" is applied per 10^6 tons (Fig. 5.1a). As a result, this objective gets a higher priority and the production scheduling of lignite having a better quality but with an insufficient amount is being avoided.

The optimizer is applied to a selection of the resource models defined in the previous section. In total, there are three different optimized mine plans for an optimization period of 4 weeks, August 2–29:

- Optimized mining plan achieved by applying the optimizer to the prior resource model of Case 1.
- Optimized mining plan achieved by applying the optimizer to the posterior resource model of Case 1 (which is updated between July 19 and August 4).
- Optimized mining plan achieved by applying the optimizer to the benchmark resource model of Case 1 (which is updated between July19 and August 13).

Next, these obtained best schedules are applied to the benchmark resource model (Fig. 5.1). This is done in order to see the improvements during the mining operations, when using the prior model, the most current model and the benchmark model. Similar to the weekly job scheduling mine planning experiments, the benchmark model is assumed to be the reality since it is the most up-to-date, and therefore the most precise resource model. An approximation of VOPI is calculated between the benchmark and the prior model, whereas the VOI is calculated between the posterior model (August 4th model; updated between July 19 and August 4) and the prior model.

Figure 5.13 shows the optimized mining sequences for the three different mining plans calculated by the optimizer. It is expected to observe more differences between the mining plans created using the prior and the benchmark resource model than between the mining plans created using the benchmark and the posterior (updated till August 4th) resource model. A difference in the mining sequence means that either a block is scheduled for mining in different extraction periods in the two mining plans in question or that a block is scheduled for mining in on mining plan but not in the other. Figure 5.14 presents the lignite's ash content for each time period based on the benchmark model (most accurate model), when applied to optimized mine plans explained in Fig. 5.12. Following observations can be derived when comparing the updated model to the prior model:

- a better fitting of the average ash values to the target of 10.7%,
- a better fitting of the average ash values to the target value area (defined from 0–15%), and
- a decrease in the uncertainty range.

To quantify these findings and the VOI, penalties for deviating from the target ash value are used (Fig. 5.15).

The darkest column represents the calculated penalties for the best mining plan, which is optimized based on the prior resource model and later this mining plan is

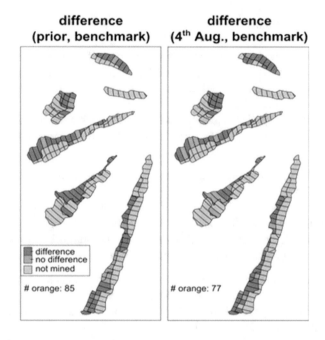

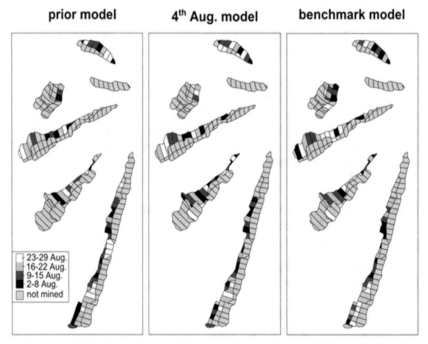

Fig. 5.13 Illustration of the optimized mining sequences and differences to the benchmark model

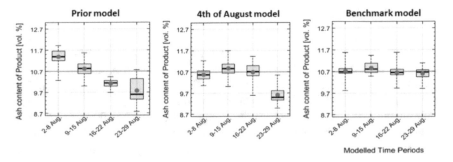

Fig. 5.14 Box plots for quality ash of the weekly scheduled lignite

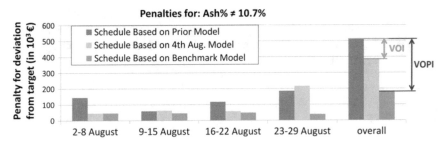

Fig. 5.15 Cost calculations of deviating from the target quality (ash%)

applied to the benchmark resource model. The lightest column represents the calculated penalties for the best mining plan which is optimized based on the posterior resource model (August 4th) of that case and later this mining plan is applied to the benchmark resource model. The medium darkest column represents the calculated penalties for the best mining plan, which is optimized by using the benchmark resource model and later this mining plan is applied to the benchmark resource model. A decrease in penalties is expected, when the optimized mining plan is calculated based on a more accurate/up-to-date resource model. When looking at each time step individually, this is not always the case. Since the optimizer calculates an overall best mining plan, single time steps can show different behavior for the penalty than the overall penalty of the entire optimization. For the overall penalty, the expected behavior can be observed: The VOI is 124.11 T€, whereas the VOPI is 327.09 T€. These significant benefits of the resource model updating framework are achieved for only one quality parameter, for a very short optimization time frame (in comparison to the life of mine) with limited localized updates. Therefore, annual cost reductions or savings between € 1.1 Mio and € 1.4 Mio could be approximated. For other parameters, like the calorific value or sulfur content, which also have a major impact on the lignite quality and consequently the revenues, similar calculations can be made.

5.4.3 Conclusions

Using the resource model updating framework in combination with the mine optimizer, the performed case study proves that the deviations from the defined target quality are reduced. In summary, for weekly job scheduling, the calculated VOI varies between € 2,000 and € 8,000 for a five-day mining period. For short-term mine planning, the calculated VOI varies from € 92,000 to € 124,000 for four weeks of mining. These numbers will lead to an approximately € 100,000–€ 550,000 annual cost reduction or saving using the weekly job scheduling optimization model. Using the short-term mine planning optimization model, which optimizes the extraction sequence, even higher annual cost reductions or savings could be possible—€ 1.1 Mio–€ 1.4 Mio.

Considering that these savings can be achieved by only using available data from production monitoring, without the requirement of new hardware installation or analysis costs, the payback of implementing a closed-loop approach in any operation will be achieved within few months.

References

R.J. Barnes, Cost of risk and the value of information in scanning, in *Application of Computers and Operations Research in the Mineral Industry* (Society of Mining Engineers of AIME, 1986), pp. 459–469

E.G.D. Barros, P.M.J. Van den Hof, J.D. Jansen, Value of information in closed-loop reservoir management. Comput. Geosci. **20**(3), 737–749 (2016)

J. Benndorf, C. Yueksel, M.S. Shishvan, H. Rosenberg, T. Thielemann, R. Mittmann, O. Lohsträter, M. Lindig, C. Minnecker, R. Donner, W. Naworyta, RTRO–coal: real-time resource-reconciliation and optimization for exploitation of coal deposits. Minerals **5**, 546–569 (2015)

R.B. Bratvold, J.E. Bickel, H.P. Lohne, Value of information in the oil and gas industry: past, present, and future. SPE Reserv. Eval. Eng. **12**(04), 630–638 (2009)

J. Eidsvik, S.L. Ellefmo, The value of information in mineral exploration within a multi-gaussian framework. Math. Geosci. **45**(7), 777–798 (2013)

J. Eidsvik, T. Mukerji, D. Bhattacharjya, Value of information in the earth sciences: integrating spatial modeling and decision analysis (Cambridge University Press, 2015)

R.A. Howard, Information value theory. IEEE Trans. Syst. Sci. Cybern. **2**(1), 22–26 (1966)

H. Mollema, Investigation into simulation based optimization of a continuous mining operation. MSc-thesis. Department of Geoscience and Engineering, Delft University of Technology. Electronic version of this thesis is available at http://repository.tudelft.nl/ (2015)

J. Peck, J. Gray, Mining in the 21st century using information technology. CIM Bull. **92**(1032), 56–59 (1999)

J. Phillips, A.M. Newman, M.R. Walls, Utilizing a value of information framework to improve ore collection and classification procedures. Eng. Econ. **54**(1), 50–74 (2009)

C. Yüksel, J. Benndorf, M. Lindig, O. Lohsträter, Updating the coal quality parameters in multiple production benches based on combined material measurement: a full case study. Int. J. Coal Sci. Technol. **4**(2), 159–171 (2017)

C. Yüksel, C. Minnecker, M.S. Shishvan, J. Benndorf, M. Buxton, Value of information introduced by a resource model updating framework. Math. Geosci. **51**(7), 925–943 (2019)

W.L. Winston, J.B. Goldberg, Operations research: applications and algorithms, vol. 3 (Thomson Brooks/Cole, Belmont, 2004)

Index

Printed in the United States
By Bookmasters